THE UNKNOWN FACTOR

BY DR . MITCHELL ALBERT WICK

COPYRIGHT 2023 BY MITCHELL WICK

COMPENDIUM PREFACE

The text introduced in this author's book is a compilation of all the new views

introduced by this author regarding the state of modern physics .This text includes

information from the following books: "Mega-physics ,A New Look at the

Universe","the Equation of Everything" ,What is the Dimension of

Time?',Megaphysics II,An Explanation of Nature",Mega-physicsIII; Nothing Doesn't

Exist", The Nth Power',Space-time ,Mass ,God and the First Event ",A Smorgasbord

of Insprations" and "What is Reality?"

The above mentioned books contain a great deal of math and while designed for

Physicists and Physics Graduate students would also be of interest to the general

public.

It discusses the Equation of Everything with it's 524,288 permutations as the

number of equation in nature,the mathematical proof of the existence of

God(Ontologic Proof),the total space-time curvature of this universe as 5×10^{-29}

radians,while a zero energy ground state is impossible and why the ground state energy must be the cosmologic constant as well as a detailed description of the First

Event with a 99 percent probability of why it is correct as well as what the function

of Dark Matter is,what caused the Big Bang,what Dark Energy is and the end result

of out universe as either "Heat Death "or a "Big Crunch". It is the purpose of this

author to tantillize young people and old to whet their appetites for advanced

learning so our civilization isn't lost when our planet becomes uninhabitable and

passes the "Great Filter"which states that a civilization will be able to colonize

another celestial body before destroying itself or being destroyed by a natural

cataclysm so that ou combined knowledge doesn't simply become a layer in dust for

an alien civilization to find and discover as undecipherable.

THE LEGRANDIAN EQUATION ; PROBLEMS WITH THE STANDARD MODEL

The Legrandian Equation is as follows: $\mathcal{L} = -\frac{1}{4} F \mu V^{\mu V} + i\psi D\psi + \hbar c + i\psi\, jY\, jk\, \psi\, k\phi + \hbar c + ||D\, u\, \Phi||^2 - V(\phi)$

Basically it states that the Legrandian=kinetic energy – potential energy

In this case potential energy is
$V(\phi)$. *The action (s) of the Legrandian over time is as follows* : $s = \int_{t1}^{t2} \mathcal{L}\ dt$

Of course the action formula in tensors is s=-1/2k^2$\sqrt{-g}$ R.

There are two distinct problems with the Legrandian Equation and the Standard

Model : 1)they do not take Dark Energy or the Cosmologic Constant into account

and 2)Gravity and anti-gravity are EFFECTS not forces.

$F\mu v$ *relates to photons in space* –
time and force symmetries. $i\psi D\psi$ *relates to fermionic field densities.* $|Du\phi|$ *relates to interacti*

Fermions including entanglement and boson interactions. Electromagnetism is
$D=\partial\Phi - ig\, cQA\, \psi$ *whereby* $D =$
$\partial\mu$ *with particles of two types of spin with fermions having* $\frac{1}{2}$ *integer spins and bosons integer*

Spins. The kinetic energy component discusses how bosons will behave
as $\partial\, uB$ *and the strong force is* is g s. and how it relates to fermionic field strength.

Space-time curvature is R in the equation $8\pi\, T\, ab = 1/R^2$ and R in the action
formula -1/2k^2$\sqrt{-g}$ R. It also relates in the line element as ds^2=dx2+dy2+dz2-c2dt2+dr2

Where r is the curvature metric curving space-time ds2. Dark Energy flattens space-

time curving it outward while gravity curves space-time inward. One must eliminate

the gravity component of the Legrandian and replace it with space-time curvature to correct for Dark Energy and showing gravity as an effect rather than a weak

force.$\mathcal{L}(x,t) = \dfrac{1}{2\rho(x,t)v^2} - \rho(x,t)\phi(x,t) - \dfrac{1}{8\pi G(\nabla\phi(x,t))^2}$. *Substitutung* $8\pi G\ T\ ab\ for$

-1/8πG and taking it's reciprocal of 1/R^2 gives space-time curvature or flattening

metric to replace gravity and anti-gravity as forces and utilize Dark Energy as the

reciprocal curvature or flattening of space-time. The expression 1/R2

$=8\pi T\ ab\ becomes \qquad R^2 = \dfrac{1}{8\pi}T\ ab\ \ or\ R = \dfrac{\sqrt{\frac{1}{8\pi}T\ ab\ or\ (\ 8\pi\ T\ ab)}^{\ -1}}{2}\ \ .8\pi\ T\ ab)^\wedge -$

1/2 substitutes for the gravity component of the Legranian or $8\pi\ T\ ab^\wedge -$

$1/2 - 8\pi G\ becomes\ 1/(-8\pi G)\sqrt{8\pi T\ ab\ or\ -8\pi G\ T\ ab)}$ So you get $8\dfrac{\pi G}{-8\pi G} -$

$T\dfrac{ab^3}{2} \qquad [(\nabla\phi(x,t)]^\wedge 2.\ \ 8\pi\ T\ ab\ \dfrac{-3}{2}\ or\ 8\pi G/\sqrt{8\pi T\ ab\ or\ \ 8\pi G\ T\ ab)}^\wedge 1/2.$

where T ab as the stress energy tensor between inertial mass and gravity curving or

flattening space-time is the energy density of

matter

ρ for any observable or metric g acting at time t with v being potential energy

ϕ being the fermionic field density and nabla ∇ being the graviton field

curving space-time as the mass component of the energy density of

matter.Substitute-1/T a b to be R a b or in this case R(x ,t) and this will correct

gravity and Dark Energy with space-time curvature and reciprocal of the square of

curvature as R and 1/R ^2where 1/R2= $\sqrt{\dfrac{8\pi G}{c4}}$ $T\ ab8\dfrac{\pi G}{c4}\ T\ ab\ \ and - 1T\ ab$ for

gravity. The expression -1/8πG *is substituted by* $8\pi G\ T\ ab)^\wedge -1/2\ or - 8\pi G/$

$8\pi G\ T a\ b)1/2\ or\ -\ 1\ \sqrt{T\ ab}$ in other words the square root of the stress energy tensor(-1) or e^iπ $(Euler's Identity)$ makes it e^iπ $(T\ ab)^1/2$.

HOW THE LEGRANDIAN EQUATION FITS INTO THE EQUATION OF EVERYTHING

SPACETIME=GROUND STATE ENERGY LEVEL + or – space/mass

The region of space-time acts upon and is acted upon by any and all energy levels

with recombination. The tensor of the fourth degree with its 256 permutations as in

the matricies and or determinants. This reflects the derivative of the tensors of the

fourth degree acting or being acted upon by all energy levels from the ground state

or cosmologic constant to the maximum free energy which must be applied to the

Legrandian expression

$\mathcal{L} =$

Kinetic energy $-$

Potential energy where the maximum free energy is the net difference between kineitc and p

Potential energy or the Legrandian. In the case of the ground state the kinetic

energy is the cosmologic constant with its anti-gravitational effect and there the

potential energy approaches 100 percent for each plane being separated from other

planes. The maximum free energy is kinetic as Dark Energy pushing apart all matter

into fermions and eventually pushing apart space-time into a Big Crunch. In this

case potential energy approaches zero and kinetic energy approaches 100 percent.

Γ*(derivative of a tensor)for the region of topological space being dilated or constricted*

By the two dimensions of time (one flat being superimposed on one spiral space-

time continuum with a finite number of universes. The location of our universe is

based on the spring effect where the string constant acts with a bow being pulled back being the Time Oscillation Paradox to a point of maximum potential energy

and the release of the bow or spring being kinetic energy being slowed or resisted

by the resisitance of space over spiral time turning space-time spiral. The velocity of

space-time is

πc so the recoil velocity of our universe is limited by kx where x is the displacement

and y is the distance of that universe with each component of spiral space-time. This

occurs throughout all 360 degrees or 2π radians with the net displacement being

$2\pi \sin \theta$ which is the forward component of the net displacement.

$$\Gamma(R\ abcd^{cdra} \ldots Rdcra^{abcd})\ over\ \Lambda\ to\ \mu(manimem\ free\ energy)$$

With entanglement being F uv^ uv from the Legrandian Equation acting with space-

time curvature being

1/R(curvature)=$8\pi\ T\ ab$ where T is the stress energy tensor.

$$\Gamma\ abcd\left(\frac{1}{\sqrt{N)}-\frac{1}{4}F\ uv^{uv}+i\psi\phi\psi} + hc + ||D\mu\phi^2||\ where\ F\ is\ force\ symmetries\right.$$

And $|D\mu\phi|$ is the average of the determinant of space acting on time.

Of course this equals

$$\Lambda + or - 2\pi c^2 \left[R\ abc+or-\tfrac{1}{2}R\ g\ ab\right] \div\ i\hbar\rho\ ab\ for\ the\ Ricci\ Tensor$$

R a b indicating inertial mass equaling $i\hbar\rho\ a\ b\ from\ Poisson's\ equation$ the

derivative operator of a two component energy system equals 4 pi

$\rho\ where\ \rho\ is\ the\ energy\ density\ of\ matter. \hbar =$

$h\frac{}{2\pi}$ where 2π goes in the numerator along with c^2 from $\rho = R\ ab/c^2$

POISSON's EQUATION

The derivative operator of any dual vector field
$=4\pi\rho$ *whereby ρ is the energy density of matter*

1. Time occupies two disparate dimensions according to recent studies as mentioned previously. Also since the new measure of time is not via clock but by the number of rotations or interactions of a subatomic particle such as the electron or fermions with electrons orbiting a nucleus each orbit can be measured and the number of orbits per millisecond can now be determined. This will clearly illustrate that as a heavy mass is approached time dilates or becomes slower with reference to the observer and this dilation of time constricts space (Wick's Law of Universal Mass). Also as a heavy mass is left and the observer is moving toward a lighter mass or a near vacuum of space time speeds up relative to the observer space is dilated or flattened as it approaches flat space-time of the near vacuum of deep space.

Applying Poisson's Equation to space-time where space-time and inertial mass are the two derivative operators and with time being both flat and spiral according to recent experiments the derivitive operator is designated by the Christoffel

SymbolΓ $\Gamma space - time, mass = 4\pi\rho(space - time) + 4\pi\rho(inertial\ mass)$

Taking out
π *leaves π $4\rho + 4\rho$ where inertial mass acts upon and is acted upon by*

space-time yielding $8\pi T\ ab$ where T a b is stress energy between inertia and the effect of gravity.

As anti-gravity curves space-time outward and gravity curves space-time inward the flattening of space-time with the push of anti-gravity as in Dark Energy or the Cosmologic Constant in reality besides gravity being an effect where mass is the cause and curving space-time is the action from mass causing the effect of gravity anti-gravity is still gravity but because the curvature of space-time here is negative or flattening or reciprocal curvature the only way this effect can occur is if the cause mass is the reciprocal of mass which causes space-time to curve in. Anti-matter appears to fulfill that goal with WHITE HOLES which repel light (reflect) and do not allow matter to enter as the repulsive effect of anti-gravity will spill the matter into flat space but as in a black hole gravity will absorb the matter and absorb light. Light is bent by gravity indicating that photons have a mass of 5x10^18 kV/c2.

THE FINE STRUCTURE CONSTANT AND ENTANGLEMENT

The fine structure constant designated

by $\propto$

$(alpha)$ is $\frac{1}{137}$ $without$ ant $units$ of $measurement.$ $Entanglement$ in $terms$ of $qu-$

$bids$ is $1/\sqrt{2}$ for $qubids$ 1 and 0 and   for the N-bid reflecting N energy states

or eigen-states it is

$$1/\sqrt{N}\,......1/\sqrt{2} = \frac{1}{1.414} = 0.707 \; for \; the \; binary \; qubits \; of \; 1 \; and \; 0.$$

$$1/\sqrt{\infty} = 0. \quad mu(\mu) which \; is \; the \; maxiumum \; free \; energy \; state \; is \; approximately$$

10^77 joules so the entanglement is approximately $1/\sqrt{10^{77}} = \frac{1}{10^{38.5}}$

1/137 reflects the infinite sum from the cosmologic constant to the maximum free

energy with both positive and negative numbers with 1/137 being the central

inflection point although negative energy is yet to be proven as it would be an

energy vacuum as in a Big Crunch where space-time flushes down to a point. It is

possible using a sine wave diagram the x axis would intersect at the fine structure

constant with N being the ordinate or y axis as the energy state(s) in question and x

or the abscissa being the degree of entanglement with the 0,0 point being 1/137 as

the z axis. This is more easily ascertained by a diagram but shows the fine structure

constant relates to eigen-states of energy with regard to entanglement which on a

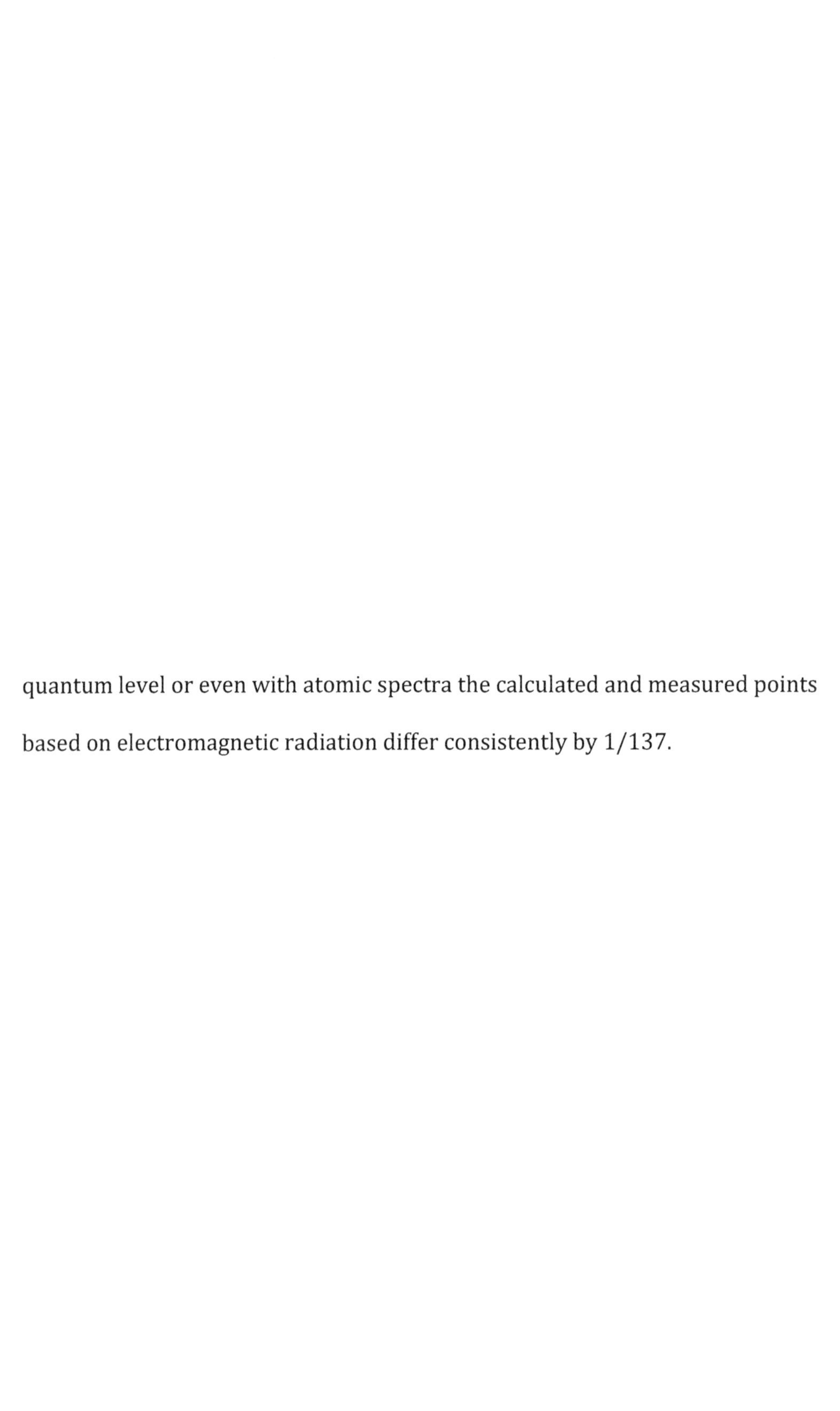

quantum level or even with atomic spectra the calculated and measured points

based on electromagnetic radiation differ consistently by 1/137.

HOW WICK'S UNIVERSAL LAW OF MASS CAN BE SHOWN HOW TO EXCEED THE SPEED OF LIGHT

According to the Lorenzian Transformations as one approaches the speed of light inertial mass approaches infinity and time dilates to where it almost stops. In every active black hole whether it contains dark matter or not time dilates and space constricts but not entirely. In the case of an observer approaching the speed of light length shorts ,time dilates or becomes infinitely slow and space constricts toward zero. Without space the observer traveling at what approaches "c" hits a brick wall when it come to space as infinite inertial mass means curvature of space-time approaches infinite curvature which can only occur with a point which is what space will become according to this author's Universal Law of Mass. How does one exceed the speed of light or 3x10^8meters/sec? Increase the space the observer or object or photons are traveling. Without space one cannot accelerate. Create space. This can be done by either creating a vacuum or near vacuum or flatten space-time to the degree as it does in a near vacuum of deep space. Dark Energy is anti-gravitational and anti-gravity curves space-time outwardly while gravity curves space-time inwardly. As a push in the engines with Dark Energy one can breech the speed of light and exceed it as space-time is UNCURVED and FLATTENED OUT OPENING IT UP FOR THE EGRESS OF THE OBSERVER AND ALLOWING THE OBSERVER TO ACCELERATE PAST "c" or the speed of light. Another way to breech the speed of light is by creating a vacuum egress which would suck a hole "in the brick wall "with space. VACUUM ENERGY WOULD ALSO PULL THE OBSERVER PAST THE SPEED OF LIGHT. It has been mentioned previously by this author that space-time travels at velocities fast than the speed of light as do tachyons. Apparently tachyons can create space by either creating a vacuum state or dark energy which is the cosmologic constant times infinity.

 Please note though that when photons or electromagnetic radiation is measured at 2.99×10^8 meters/sec in a vacuum photons have no clear resting mass and measured mass as 5×10^{-18} kV/c2 which is miniscule rather than the indescribable mass as in the Lorenzian Transformations for matter approaching the speed of light with a progressively infinite approaching mass(inertial mass).In this case the event horizons of active black holes mimic matter approaching the speed of light more than light itself .Of course in this case space-time spirals or flushes into a black hole with dilated(slow) time constricting space into a nearly asymptotic cone. In this case also energy and matter are treated as two separate entities rather than being interchangeable as in $E=\sim mc^2$. To summarize matter engaged in the Lorenzian Transformation with an approaching infinite inertial mass at v~c does not include light or electromagnetic radiation where mass approaches zero at v=c. Recall also that depending on the medium it travels through light travels at all speeds except for zero which was brought out by this author in previous books.

DOES ENERGY CURVE SPACETIME?

Energy does curve space-time because photons have a miniscule mass of 5.5x10^-18kV/c2. The curvature of space-time by energy is extremely slight which is why the universe is almost flat. The energy from "The Big Bang" equivalent to the background microwave radiation heat space by 2.74 degrees kelvin and curves space-time only 5.5x10^-29 radians as the void of deep space comprises 99.999% of our universe. A vast majority of space-time curvature occurs at or near the event horizon of any active black hole with the remainder being near stars and nebulae.

Space-time curvature can be calculated as the reciprocal of the Einstein Constant or Gravitational coupling constant times the stress energy tensor measuring stress between inertia and gravity such that $8\frac{\pi G}{c4}\ T\ ab = (R\ ab)^{-1}\ or \frac{1}{R}$ where R ab is the curvature of space-time caused by the metric g ab and T ab is the stress energy tensor of the metric g ab. This 1/R can also approximate the reciprocal of the Cosmologic Constant Λ which is the anti-gravity component of Dark Energy which will eventually break apart all mass down to fermions and possibly can a "Big Rip" in space-time in approximately 22 billion years. The reciprocal of the Cosmologic Constant is approximately 10^55 joules which should approximate Dark Energy if and only if there are a FINITE number of completely parallel planes prior to the first event. The Cosmologic Constant was the weak anti-gravitational force(or effect) that prevented these planes from touching and forming dimensions and therefore if space-time is INFINITE there must have been an infinite number of parallel planes and the cosmologic constant would be multiplied by infinity making it greater than

10^55 joules resulting in a Big Rip ,Flush or Crunch for our universe rather than the

accelerated expansion of galaxies away from each other reversing resulting in "Heat

Death" or everything stopping at or near absolute zero. As the reciprocal of the

Cosmologic Constant or 10^55 joules may be the maximum energy level   designated

as

$\mu(mu)$ *with the ground state energy level being the Cosmologic Constant or* 10^{-55} *joules*

The total Free Energy of our Universe would be the mass of our universe of

10^54kg(3x10^8 meters/sec)^2   or 9 x10^70 joules while the reciprocal of the

Cosmologic Constant is 10^55 joules resulting in Dark Energy extinguishing the

accelerated expansion with 10^15 joules remaining when and if the universe

contracts resulting in Heat Death. Again this depends on whether Dark Energy is

infinite or finite.

A New Development in the Dimension of Time

Studies published in late July 2022 through the University of Southern California Physics Department on their research in discovering a new state of matter with measurements in qu-bits came about another discovery. The dimension of time in space-time is actually two dimensions; one flat in the absence of mass or it's equivalent in energy and one following the Fibonacci sequence resulting in spiral configuration. As time defines the region of space if the Fibonacci sequence is applied to time and time influences the configuration of space and therefore space-time then the configuration of space-time under certain circumstances is spiral, which is precisely what this author has been repeatedly stated since the publication of "Megaphysics; A New Look at the Universe" Dorrance Press(2003). This is the configuration of space-time forming the space-time continuum following "The First Event" as a "Time Oscillation Paradox "with a centrifuge effect forming the string dimensions near the center or apex of the spiral and macroscopic dimensions forming the multi-verse in other regions of the spiral in a slingshot effect after the spin and time oscillation. Prior to this when tachyons were only above the speed of line and fermions were only below the speed of light times' arrow went in one direction below the speed of light and the opposite direction above the speed of light. Prior to the oscillation the dimension of time was flat or devoid of curvature as there was only negligible mass due to the ground state energy level from the Cosmologic constant separating the infinite number of completely parallel planes and preventing them from touching and becoming a second dimension with time infinitely dilated or tremendously slow being the first dimension. Our universe was still formed from a' Big Bang" or Inflation but these were preceded by a "Big Crunch"

or Flush of fluid like space-time. As a result space-time is approaching asymptotic

flatness in our universe with extreme curvature primarily where mass is

concentrated such as at Black Hole event horizons. Itzhar Bars from the University

of Southern California was working on pulsed light in terms of qu-bits in order to

find a new phase of matter and the pulses followed a spiral fractal pattern in the

Fibonacci sequence utilizing a quantum computer. The Ytterbium atom was

transformed into Quantinium as was also detected by Phillip Dumitrescu at the

Flaurum Institute with computational quantum physics based on qu-bits. Searching

for the "quipu" with regard to time crystals it conclusively proves that the likelihood

of the First Event was this author's "Time Oscillation Paradox" rather than "The Big

Bang "or Inflation which occurred later. Atoms struck with a pulse laser in the

Fibonacci sequence revealed the extra (spiral) dimension of time which existed after

the First Event from the Time Oscillation as indicated in the article by Dr. Zeeya

Merali in Scientific American and Live Science.

      The Fibonacci Sequences is based on *psi*φ *whereby* $\varphi =$

$$1 + \sqrt{\frac{5}{2}}$$

and 1.6177 *is approached via the sequence* $1\ \frac{1}{1} = 1 \quad 2\ \frac{2}{1} = 2 \quad 3\ \frac{3}{2} = 1.5 \quad 5\ \frac{5}{3} = 1.67$

And psi is approached. The Golden Ratio with respect to time can be incorporated

into space-time as the derivative operator of the tensor of the fourth degree

becomes an infinite sum with regard to the Fibonacci numbers.

Γ *abcd as an infinite sum* $\sum_{\Lambda}^{\mu} 1,2,3,5,8,13 \ldots \ldots \varphi$ *or* 1.6177 *relating to* $1 + \sqrt{5} \div 2.$ This refers to the energy levels from the ground state or Cosmologic Constant to mu the maximum free energy relating to potential energy of the space-time continuum. Space-time is acted upon by all these energy levels with each energy level being multiplied by Psi as an infinite product of the infinite sum of the Fibonacci numbers and the derivative of these is multiplied by

$1/\sqrt{N}$ $\;$ for N qu − bits to incorporate entanglment.

Γ abcd $1/\sqrt{N}$ $\;$ + φ |

Λ to μ =

ds^2 $\qquad$ The derivative operator of the tensor of th fourth degree of space −
time incorporating entanglement as 1/

$\sqrt{N}$ $\quad$ with qubits along all the energy states or eigenstates plus φ(psi) $+\dfrac{1}{\varphi}$ as an infinite sum

Incorporates the Fibonacci sequence for the golden ratio relating to spiral space-time+

$\prod_{\Lambda}^{\mu} ds^2$ $\qquad$ repressents the two dimensions of time acting upon space across all the masses

And their energy equivalents from the ground state or cosmologic constant to the

maximum excited free energy state of mu with entanglement incorporated. This

massive equation describes space-time equaling Λ + or − $2\pi c\^2[R$ abc + or −

$\dfrac{1}{2}R\,g$ ab $\quad \div \quad$ i h (ρ abcd). Here $2\pi c^2$ comes from $\hbar = h\dfrac{}{2\pi}$ and R ab =

i$\hbar\rho$ ab $\div$ c^2 as an approximation of e = mc^2.

A NEW DERIVATION OF SPACETIME=SPACE/MASS

In terms of metric tensors space-time is curved Lorenzian Space-time or R(region

of topological space acting upon or being acted upon by the tensors of the 256

permutations of space-time in tensor of the fourth degree represented as R abcd

with the different permutations of the different eigen-state of energy from the

ground state or the cosmologic constant to 10^77 joules tying in with the energy

associated with the Higgs Field and the Grand Unification Energy of 10^19 giga

electron volts with recombination from each energy level commixing with each

other energy level.this is entanglement.

$$\Gamma R\,abcd(1/\sqrt{n}\quad 10\text{\textasciicircum}77 \text{ joules}|\text{Cosmologic Constant}||1\ \ 0 >$$
$$\mu \;=\; \Lambda + or - \quad R\,abc + or - \tfrac{1}{2}R\,g^{ab} \div \; i\,\hbar\,\rho\,abc$$

1. Here the derivative of the tensor of the fourth degree with entanglement

   over n eigenstates of energy grom the cosmologic constant to the Grand

   Unification Energy or 10^77 joules or both is spacetime and this equals the

   cosmologic constant plus Euclidian flat space + antigravity(1/2) –

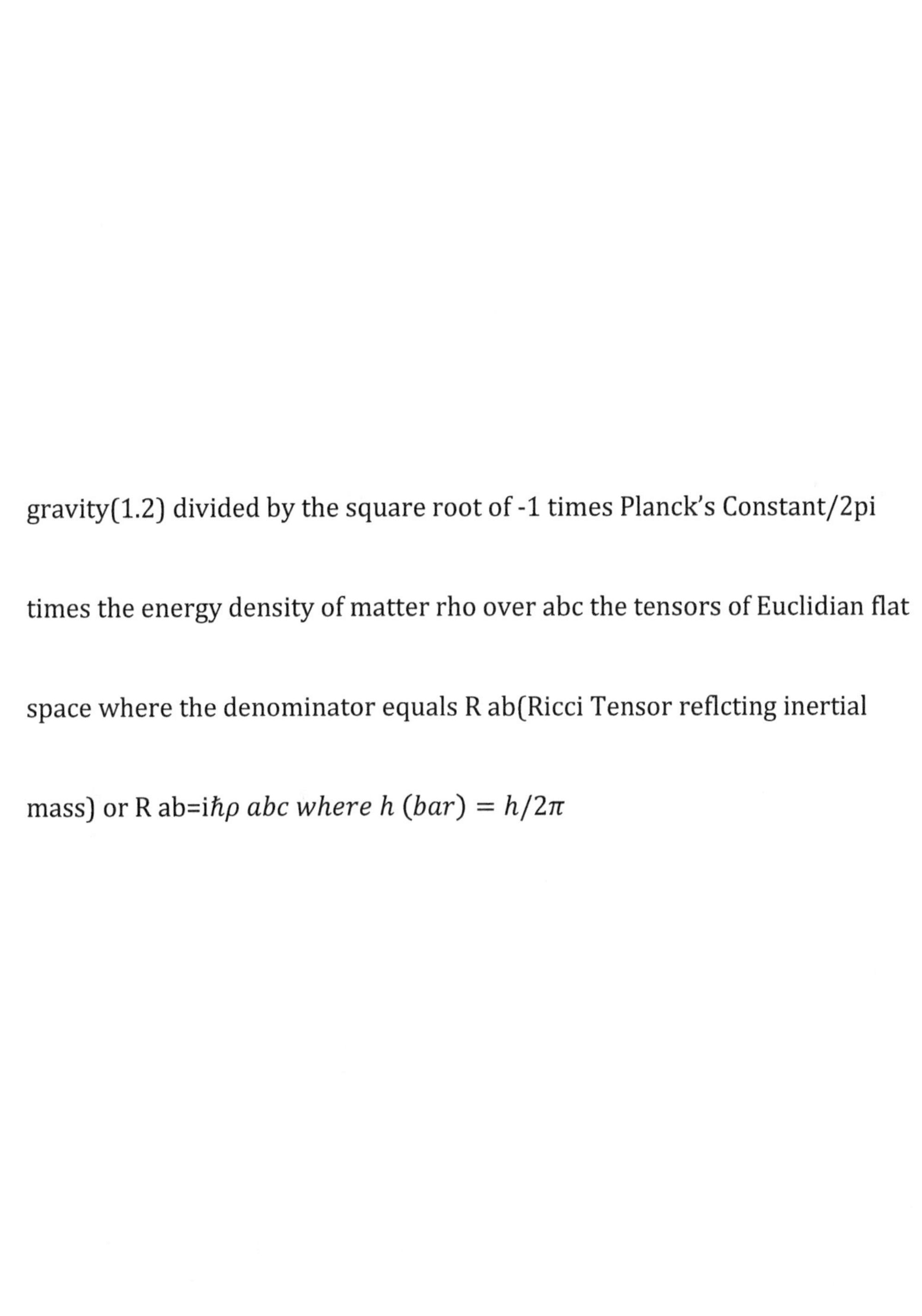

gravity(1.2) divided by the square root of -1 times Planck's Constant/2pi

times the energy density of matter rho over abc the tensors of Euclidian flat

space where the denominator equals R ab(Ricci Tensor reflcting inertial

mass) or R ab=i$\hbar\rho$ abc where h (bar) = $h/2\pi$

COMPENDIUM PREFACE

The text introduced in this author's book is a compilation of all the new views introduced by this author regarding the state of modern physics .This text includes information from the following books: "Mega-physics ,A New Look at the Universe","the Equation of Everything" ,What is the Dimension of Time?',Megaphysics II,An Explanation of Nature",Mega-physicsIII; Nothing Doesn't Exist", The Nth Power',Space-time ,Mass ,God and the First Event ",A Smorgasbord of Insprations" and "What is Reality?"

The above mentioned books contain a great deal of math and while designed for Physicists and Physics Graduate students would also be of interest to the general public.

It discusses the Equation of Everything with it's 524,288 permutations as the number of equation in nature,the mathematical proof of the existence of God(Ontologic Proof),the total space-time curvature of this universe as 5×10^{-29} radians,while a zero energy ground state is impossible and why the ground state energy must be the cosmologic constant as well as a detailed description of the First

Event with a 99 percent probability of why it is correct as well as what the function

of Dark Matter is,what caused the Big Bang,what Dark Energy is and the end result

of out universe as either "Heat Death "or a "Big Crunch". It is the purpose of this

author to tantillize young people and old to whet their appetities for advanced

learning so our civilization isn't lost when our planet becomes uninhabitable and

passes the "Great Filter"which states that a civilization will be able to colonize

another celestial body before destroying itself or being destroyed by a natural

cataclysm so that ou combined knowledge doesn't simply become a layer in dust for

an alien civilization to find and discover as undecipherable.

AN ABBREVIATED FORM OF THE EQUATION OF EVERYTHING CHAPTER SIX(6)

If one hits the jackpot a feeling of euphoria permeates the body. This happened to

this author twice in the realm of Physics. The first time in 2002 when du/u=lnu was

determined to be the famous zero in the quantum ground state in a relativistic

universe. Although incomplete as the quantum ground state energy level is the

Cosmologic Constant and not zero it was only an approximation. However in

June,2021 that same feeling of euphoria came about as M Theory(membrane

theory) was determined as "The Theory of Everything"that Dr.Michio Kaku was

looking for. This is because the equation of a circle circumference=2(pi)r fits very

neatly into the one inch equation with the circumference being space-time and the

radii equaling the sum total of all the Riemann Forces of Nature both positive and

negative totaling

$\mu(mu)$ *whereby the equation the mass of a string =*

2π*(tension of each string)(number of strings) =*

total energy equalling $10^{77\,joules}$ *or* μ*(free energy of all the phenomena of nature the maximu*

Energy state or eigen-state (quantum state).Poisson's equation is $2(2\pi\rho = \frac{2\mu}{2} =$
μ *therefore* $\mu =$
ρ *or the maximum energy density of matter and the mass of a string (type IIA closed)relates*

$$\Gamma^{-1}\ abcd\ [1/\sqrt{E(n)}\ |\ \Lambda\ to\ \mu\ =\ E(0)\ + or - \ 2\pi\ c^2\ [\ R\ abc + or - \tfrac{1}{2} R\ g\ ab \div ih\rho\ abc$$

In essence the anti-derivative of the tensor of the fourth degree representing curved

space-time fro the cosmologic constant to mu as limits of the anti-derviative as

represented by the Christoffel symbol to the -1 power with q-bits represented by

1/the square root of all the energy states or eigen-states representing matter gives

each and every perturbation of space-time caused by or being the cause of the

space-time curvature metric representing gravity and anti-gravity but also the

Strong Force and Electromagnetism must play in with entanglement.

Electromagnetic moments occurred with the spin momentum vector of the first

event with the Time Oscillation Paradox and correspond to the 1-brane and 2-brane

levels in M Theory. In terms of eigen-states the eigen-states representing the energy

equivalent of the mass of an electron changing energy levels according to the Bohr

energy states 1s^2,2s^2,p^6.... And the Strong Force relating to E=mc2 where is

$R\ ij = i\hbar\rho ij/c2$

Antiparticles, Dark Energy ,and the Big Bang   by Dr. Mitchell Albert Wick

Antimatter was first discovered in 1932 and since then myriad antiparticles have

been discovered including the positron, anti protons, and antiquarks. Antiparticles

are being isolated at Cern, Switzerland in the Hadron Collider but as of this date a

mass of the antiparticle has not been conclusively discovered as positive. It has been

noted that antiparticles have the opposite charge to matter particles but it has not

yet been ascertained that antiparticles have a positive gravity. If antiparticles

mutually repel with antigravity instead of attract with gravity a plausible

mechanism for the 'Big Bang' can be made for the quantum bubble, At Planck Time

10-43 seconds a 50:50 mix of matter and antimatter caused a symmetrical 360

degree(or 2 pi radians)orb blast because of the mutually repulsive force of

antiparticles which forced the explosion converting 99.9999% of the antimatter into

Dark Energy via the formula E=mc2 postulated by Einstein where m=mass of the

antimatter. Whether the mass is positive or negative is up to speculation but since

the Law of Conservation of Energy was purportedly violated by the 'Big Bang' with

antimatter having a positive mass it can be concluded by induction that the mass is negative which is why so little antimatter exists today. As the antiparticles repel

dark energy pushed outward in all directions carrying the balance of matter with it

which it subsequently attracted to other matter by gravity but   (kl)not

to antimatter. The formula (my derivation is included in this work)R jikl- Rji

=-8p[(Geji)=g ji(in four dimensions of space-time reflects the magnitude of dark

energy's repulsive force as curvature of space-time(opposite curvature to

gravity)(gravity was described by Einstein as space-time curved by mass)and the

right side reflects the mutually repulsive force of antiparticles and g ji is the metric

of the antiparticle where i=initial event and j=final event.- 8 pi G where G=the

gravitational constant is from the Cosmologic Constant of Albert Einstein which

reflects the mutually expanding repulsive force pushing galaxies father and farther

apart purportedly fro Dark Energy e=2.71828 and e ji is the vector product of e

from j to i.

This assumes a symmetrical 360 degree Orb Blast in the 'Big Bang' and R is the anti-

gravitational effect on particles and R ji is the antigravity effect on antiparticles. In a symmetric orb blast of 360 degrees or 2 pi radians there is an angle of trajectory of 180 degrees or pi radians. The cosine of pi radians=-1 which explains the -8(pi)G on the right side of the formula. There are an infinite number of 180 degree slices in a perfect sphere so the angle of trajectory for an isotropic universe must be 180 degrees or pi radians.

Above formula   Rjikl-Rji=-8(pi)G e ji=g ji or -8(pi)G e(ij,ji)= g ji or R jikl-Rji =-8(pi)G e(ij,ji)=g ji

1/Rij^2 =space-time of the cosmologic constant or Dark Energy Rij'R ji^2 as the vector product reveals e ji R2 ij R ji/THE ABSOLUTE VALUE OF R SQUARED ij times the absolute value of R ji where R 2ij R ji/the absolute value of R SQUARED ij times the absolute value of R ji =cos theta and theta is pi radians domain 0, or equal to theta , or equal to pi range cos theta -1 , or equal to cos theta , or equal to 1 theta is pi radians. As the trajectory for an isotropic universe is pi radians an infinite number of 180 degree slices are in a perfect sphere The cosine of pi=-1 g ji=metric for an antiparticle g ij=metric for the matter particle. R ji=1/R ij 2=1/8(pi)G .g ji where R ji is antigravity for the antiparticle1/ R ij squared is spacetime of the cosmoloigic

constant 1/ 8(pi)G   is the cosmologic constant and g ji is the metric of the anti-

particle.The vector product R ij'R ji=e (ij,ji)=cos (pi)= - e(ij,ji) R jikl-R ji kl=-e(ij,ji)= g

ji/8(pi)G where R jikl= covaruiant t ensor for space-time curvature from

antiparticle of metric g ji on antiparticles R kl=contravariant tensor kl with

covariant tensor ji ji from antiparticle g ji kl is from gravity of contravariant tensor

from metric g ij for matter.   K

                ji

        ijkl
Note R =antigravity affect on particles from antiparticles , The -8(pi)G reflects the

anti-gravitational moment of antiparticle anti-particle interaction

        kl
Rjikl-Rji=-e(ij,ji)=-8(pi)G e(ij,ji)=g ji   Q.E.D.

CHAPTER FIVE
THE FIRST EIGEN-STATE OR QUANTUM STATE OF ENERGY IS THE COSMOLOGIC
CONSTANT

After the first event or Time Oscillation Paradox gravity or spin 2 vector bosons

formed. This resulted from tachyons dropping to the speed of light and below the

speed of light and the spin 2 vector bosons had inertial mass so R ab>0 or positive.

This formed gravity which curved   space-time inwardly caused by the mass of the

spin2 vector bosons. Such that one t=gets -1/2 R g ab so $0 = \sim\! \Lambda \, g \, ab + R \, ab -$

$\frac{1}{2} R \, g \, ab = \frac{8\pi G}{c4} \; T \, ab.$ *The R ab is comprised of bosons, leptons and gluons.*

Regarding gravity after the First Event $-\Lambda \; g \, ab = R \, ab - \frac{1}{2} R \, g \, ab = \frac{8\pi G}{c4} \; T \, ab.$

$$\kappa \, T \, ab = \frac{1}{\sim\! R^2} = \Lambda \, g \, ab \qquad \kappa = \frac{8\pi G}{c4}$$

$$\Lambda \, g \, ab = \frac{1}{\sim\! R^2} = R \, ab + \frac{1}{2} R \, g \, ab \;\; and \; relates \; to \; antigravity$$

Stress Energy relating to antigravity is -8$\frac{\pi G}{c4}$ $T \, ab$

~R ab)^2 is the space-time curvature caused by the action of the metric g ab

Λ g ab = Quantum Ground State of Energy = ρ ab

The Λ *has a metric of g ab relating to the miniscule mass of space hadrons*

(leptons and gluons)below the speed of light and tachyons above the speed of light.

R ab^2=space-time curvature of*ρ ab = Λ g ab = ∞² = ∞*   so 1/R^2=1/∞² =

0 *or flat spacetime when the energy desnsity of a vacuum is* Λ g ab.. −Λ g ab =

~Λ g ab althoughboth approach zero as an asymptote. Λ g ab =≠ −Λ g ab = Λ g ba

The action of the metric g ab relates to the space-time curvature metric such that S=-
$$1/2\kappa2\sqrt{-g\ R\ g\ ab}\ and\ -\frac{1}{2\kappa2}\frac{g}{R}g\ ab\ where\ \frac{g}{R}g\ ab=\sqrt{-g\ R\ g\ ab}.$$

HOW THE SPACETIME CURVATURE OF ANTI-GRAVITY AND SPACE-TIME

CURVATURE FOR GRAVITY RELATES TO FLATTNESS IN THE VACUUM OF SPACE

AND THE INFINITE CURVATURE AT THE EVENT HORIZON OF A BLACK HOLE

Dark Matter causes the continuity of time or the sequencing of events. THE MASS OF

DARK MATTER PRIOR TO THE FIRST EVENT=1.16X10^-188/-1kg< mass of ordinary matter ~0. The mass of ordinary Baryonic Matter prior to the First

Event=1.04x10^-89kg as the mass of Dark Matter=mass of ordinary matter/i

DOES ENERGY CURVE SPACETIME?

Energy does curve space-time because photons have a miniscule mass of 5.5x10^-18kV/c2. The curvature of space-time by energy is extremely slight which is why the universe is almost flat. The energy from "The Big Bang" equivalent to the background microwave radiation heat space by 2.74 degrees kelvin and curves space-time only 5.5x10^-29 radians as the void of deep space comprises 99.999% of our universe. A vast majority of space-time curvature occurs at or near the event horizon of any active black hole with the remainder being near stars and nebulae.

Space-time curvature can be calculated as the reciprocal of the Einstein Constant or Gravitational coupling constant times the stress energy tensor measuring stress between inertia and gravity such that $8\frac{\pi G}{c4}\ T\ ab = (R\ ab)^{-1}\ or\ \frac{1}{R}$ where R ab is the curvature of space-time caused by the metric g ab and T ab is the stress energy tensor of the metric g ab. This 1/R can also approximate the reciprocal of the Cosmologic Constant Λ which is the anti-gravity component of Dark Energy which will eventually break apart all mass down to fermions and possibly can a "Big Rip" in space-time in approximately 22 billion years. The reciprocal of the Cosmologic Constant is approximately 10^55 joules which should approximate Dark Energy if and only if there are a FINITE number of completely parallel planes prior to the first event. The Cosmologic Constant was the weak anti-gravitational force(or effect) that prevented these planes from touching and forming dimensions and therefore if space-time is INFINITE there must have been an infinite number of parallel planes and the cosmologic constant would be multiplied by infinity making it greater than

10^55 joules resulting in a Big Rip ,Flush or Crunch for our universe rather than the accelerated expansion of galaxies away from each other reversing resulting in "Heat Death" or everything stopping at or near absolute zero. As the reciprocal of the Cosmologic Constant or 10^55 joules may be the maximum energy level   designated as

$\mu(mu)$*with the ground state energy level being the Cosmologic Constant or* $10^{-55} joules$
The total Free Energy of our Universe would be the mass of our universe of

10^54kg(3x10^8 meters/sec)^2   or 9 x10^70 joules while the reciprocal of the Cosmologic Constant is 10^55 joules resulting in Dark Energy extinguishing the accelerated expansion with 10^15 joules remaining when and if the universe contracts resulting in Heat Death. Again this depends on whether Dark Energy is infinite or finite.

DARK MATTER IS INFORMATION STORED NEUTRINOS IS INFORMATION

TRANSMITTED AND THIS RESULTS IN DARK MATTER BEING A NERVOUS SYSTEM

It was just determined by a United Kingdom physicist Dr .Melvin Vopson at the

University of Portsmouth 1 that there is evidence that Dark Matter which has been

considered "cosmic glue" is basically information. The question to this hypothesis is

this: dark matter curves space and mass curves space-time. The effect is gravity and

anti-gravity but to state that dark matter is information it indicates that

information has mass .Q bits contain information and describe energy levels

including entanglement. As they describe energy levels this implies that they have

mass and therefore can curve space-time. As neutrinos carry information this can

describe a nervous system (many diagrams of dark matter mimic a peripheral

nervous system and can be a peripheral nervous system for the abstraction of the

Higgs Field where tachyons act as a central nervous system.

1:Vopson,Dr. Melvin. University of Portsmouth

DERIVATION OF THE EQUATION OF EVERYTHING

EVERY QUANTITIY EQUALS ITSELF….THE TRANSITIVITY POSTULATE

THEREFORE

SPACE-TIME=SPACE-TIME

SPACE-TIME IS DIRECTLY PROPORTIONAL TO SPACE DUE TO THE LINE ELEMENT

ds^2=dx^2+dy^2+dz^2-c^2dt^2+dr^2

SPACE-TIME IS INVERSLEY PROPORTIOANL TO MASS

TIME DILATES OR SLOWS DOWN AS IT APPROACHES A HEAVY MASS SUCH AS A
BLACK HOLE EVENT HORIZON
DILATED TIME CONSTRICTS SPACE;THE MORE DILATED OR SLOWED TIME IS THE
MORE IT CONSTRICTS SPACE LIKE A LASSO AROUND SPACE BY ELONGATED OR
DILATED TIME
CLOCKS SPEED UP AT HIGHER ALTITUDES OR LARGER DISTANCE FROM THE
EARTH

SPACE-TIME=SPACE/MASS

In terms of tensors with g being any metric space-time is a tensor of the 4th degree
with covariant and contra-variant components.
R abcd=-R dcba anti-symmetrical tensor and Bianchi's Idenitity

R is the scalar or magnitude of the tensor in the REGION OF TOPOLOGICAL SPACE
abcd are the directions of the tensor acting upon or being acted upon by everything
interacting with space.

Space-time=R abcd=Λ $+ or -$ R abc $+$ or $- \frac{1}{2}$ R g ab $\frac{}{i}$ $\hbar$ ρ abc

The cosmologic constant is the ground state energy level R abc is Euclidian or flat
space +1/2 R g ab is anti-gravity as it curves space outward flattening space-time
and -1/2 R g ab is gravity which curves space inwardly or constricting space down
toward a pont .Gravity is the curvature of space-time caused by mass and it curves
space inwardlt while anti-gravity curves space outwardly flattening it also caused by
mass as a reciprocal curvature as designated by 1/time .The term R ab in the
denominator representing inertial mass equals $i\hbar\rho$ $where\ i = \sqrt{-1}$ h=Planck's
Constant and h(bar)=Planck's Constant divided by 2 pi. Rho is the energy density of
matter from Poisson's Equation and these equal the positive and negative Riemann

Forces of Nature with gravity and anti-gravity being the effects of space-time curvature caused by R ab or inertial mass or i h(bar) rho to form curvature and reciprocal curvature depending on whether time is in the numerator or denominator time vs. 1/time. Space-time has a derivative of the tensor of the fourth degree to show how space-time changes in all the energy states of matter from the ground state (Cosmologic Constant) to the maximum excited state or μ as a near infinite sum or $\Sigma \; \Sigma_\Lambda^\mu \, 1/\sqrt{N}$.

N IS THE NUMBER OF STATES INVOLVED AND $1/\sqrt{2}$ IN QBITS IS THE ON AND OFF SWITCH REVEALED AS 2. IN THIS CASE THE QBITS IS N ENERGY STATES INCORPORATING ENTANGLEMENT OF ONE ENERGY STATE INTO ANOTHER ENERGY STATE OR MULTIPLE STATES   AND HOW THESE ALL AFFECT SPACE-TIME CURVATURE(LEFT SIDE OF THE EQUATION. THE BOLTZMANN EQUATION INVOLVES TRANSFERRING STATES OF MATTER INTO ENERGY AND Z OR THE BOLTZMANN CONSTANT GETS INSERTED INTO N ENERGY STATES AS CONVERTED FROM MATTER.THESE ARE ALL DERIVATIVES OF THE TENSOR R abcd as signified by the Cristoffel Symbol Γ so $\Gamma(R\ abcd) \sum_\Lambda^\mu 1/\sqrt{N}$ $= \Lambda + or -$ $R\ abc + or -$ $\frac{1}{2} R\ g\ ab$ $(2\pi) \div i\ h\ \rho\ abc$ to designate space-time=space/mass. N=-kT ln Z where k=Boltzmann Constant T=temperature in degrees kelvin N is the free energy state or eigen-state of energy. Z is from the Boltzmann Equation as Z the state of matter associated with the Gibbs Free Energy which also gives the entropy(S) of the system as delta G=delta H-T delta S or$\Delta G = \Delta H - T\Delta S$.

As a result; Total lorenzian curved space-time is

$$\Gamma(R\ abcd) \sum_\Lambda^\mu 1/\sqrt{-kT \ln Z} \quad = \quad \Lambda + or - R\ abc + or - \frac{1}{2} R\ g\, \frac{ab}{i}\, \hbar\rho abc$$

Z=ne^E(n)/kT   where E(n) is the energy state of the Nth state of matter such that

free energy equals –(Boltzmann Constant)kT(degrees kelvin) times the natural log

of Z(Boltzmann Equation of Z=ne^-E(n)/kT

With the measurement of gravity waves from the LIGO project which measures

curvature changes in space-time some conclusions have recently been drawn. New

Scientist;2022.With a close measurement of the near fusion of two black holes one

can postulate space-time as being two nearly identical regions of space with a sharp

dichotomy in the central core and a phase differential or shift making one half

slightly out of phase with the other half. This is like slicing an apple in half with a

knofe and putting half of it a quarter of an inch higher than the other half. This

duplication of constricted space-time with a shifting of the phase differential makes

space appear to have a boundary or multiple gradients between the phases of space-

time in the area or region of the dichotomy. The difference in mass between the two

nearly adjacent black holes may make space-time slightly out of phase between the

regions adjacent to the event horizons as photographed by the Hubble. Also the

presence of dark matter inside or adjacent to black hole event horizons may

contribute to the dichotomy also. The new James Webb Roving Telescope can map curvatures of space-time through the measurement of the effect of gravity on the

curvatures of space-time called "gravity waves" although in actuality they are waves

of fluid spacetime just like an ocean (the Bose Sea)of Bosons(spin 2 vector bosons).

The multitude of primordial black holes and the number of extant black holes are far

greater than originally anticipated and now with the James Webb Telescope the

recently discovered white holes can be studies. As the content would repel ordinary

matter with anti-gravity it is possible that anti-matter is sequestered in these white

holes which would bounce off ordinary matter including light with anti-gravity and

it's reciprocal curvature of space-time or reflected as 1/time where time dilates

space toward asymptotic flatness rather than constricting it or curving it inward

towards a central point or nexus. Of course there is no evidence than a black hole

has ever come close to merging with a white hole as this may trigger another TIME

OSCILLATION PARADOX SIMILAR TO THAT OF THE FIRST EVENT WHICH MAY

DISPLACE THE GLUONS THAT HOLD TOGETHER THE LEPTONS IN SPACE CAUSING

AN EVENTUAL BIG CRUNCH AFTER THE TIME OSCILLATION. Of course there may

be such a solid gradient between these two regions of space-time that a false boundary may turn into a true boundary protecting the effects of anti-gravity in the

white hole from reacting with the gravity of a black hole.

Recently a study has been performed to show a matrix between light waves

and subatomic particles being held in that matrix. This could mimic Boso-

Einsteinian Condensate at super-cold temperatures as light is slowed to 36mph and

the mass of 5.5×10^{-18} eV /c2 can adhere to other mass involved with fermions and

possibly protons or neutrons.

CHAPTER XXXI(31)

The gradient is the progressive change in amplitude of a force,energy field or space-

time such that d26 or 26 dimensions represent d26/25

...d25/24...d24/23....d4/3...d3/2...d2/1 with forces or energy acting upon it or

being acted upon it it terms of strings with the mass of a string being

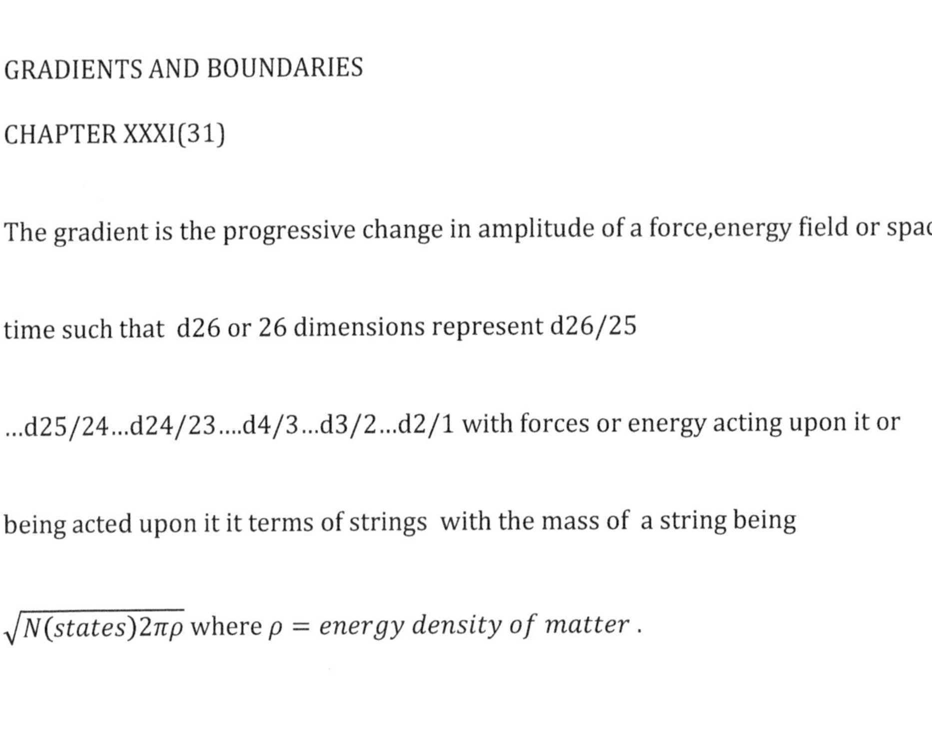

$\sqrt{N(states)2\pi\rho}$ where $\rho = energy\ density\ of\ matter$.

MEGAPHYSICS II :A DISCOURSE ON HOW SPIRAL SPACETIME LEADS TO THE
EQUATION OF EVERYTHING

BY   DR. MITCHELL ALBERT WICK
FORWARD AND INTRODUCTION
  In this author's first book "Mega physics   ,A New Look at the Universe ",it was

postulated that space time curvature followed the spiral fractal formula and was

used to locate the quantum ground state in a Relativistic Universe. This is due to the

rotational component of the open "flat" expanding universe (Friedmann Type II)

and the rotational vector referred to Godel' s Rotating Universe.

  The area of maximum rotation occurred at Planck's Time 10-43 sec and gradually

flattened out with the accelerated expansion after "The Big Bang" until it reached a

curvature equaling the space-time curvature metric of Albert Einstein based on the

Action Formula S=-1/2k2(-g)1/2R where k is the gravitational coupling constant S

is the action of any metric g ,and R is the curvature variant on space-time caused by

the metric g.

  There are two super-symmetric manifolds of space-time as dictated by Quantum

Field Theory and brought to light in this author's second book "The Equation of

Rked on the infinite dimension Hamilton-Jacobi Equations with regard to   1
viscosity solutions in an article 11.

Everything" as -1/2e-in cot theta   and -1/2 e in cot theta which correspond to l n (u)

and –l n(u) which when superimposed come to flat space time corresponding to

one. One component respponds to clockwise unwinding rotation to "The Big Swirl"

and the other to the counter-clockwise rotation of "The Big Swirl" or anti-Swirl. As

pointed out in "The Equation of Everything" theta is the angle of trajectory of "The

Big Bang" and n is the number of dimensions .e=2.71828 and i=(-1)^1/2.

BACKGROUND

In this author's first book "Mega physics ,A New Look at the Universe "the geometry of space-time was postulated as spiral. This is due to the combination of expansion and rotation where if a slice of space-time is made through an expanding and rotating object is cut and the infinite sum is calculated; it is shown that space-time is spiral and follows the spiral fractal formula of

Npi/2i$\frac{\int du}{u}$ $where\ i\ is\ the\ square\ root\ of\ -1.\ This\ also\ equals\ -\frac{1}{2e}-$
$incotangent\ theta\ as\ the\ integral\ of\ \frac{du}{u}\ is\ \ln u.\ Over\ the\ past\ eight\ decades, postulates\ have\ been$

made for the origin of this universe ,including "The Big Bang" and "Inflation

"theories .Also during this time string theory and M Theory have been devised and

evolved as the "New Physics of the 21ˢᵗ Century" .Despite this myriad attempts made

by physicists have failed to show which type of String Theory consistently explains

all phenomena in this universe. This includes the existence of six dimensional

Rked on the infinite dimension Hamilton-Jacobi Equations with regard to   2
viscosity solutions in an article 11.

curled-up manifolds called Calabi -Yau Manifolds. The hybrid of all five string

theories is M theory(Membrane Theory)which look at all five string theories from

different vantage points. An analogy is if five blind men are describing a horse from

different positions relative to the horse where all five are correct. Despite this all

descriptions of the horse are different .As a result all five descriptions are

considered dual to each other as are all five string theories. Einstein was able to

show that that space-time curvature is a function of mass as a matrix or manifold in

which an added surface or surfaces relate to a fixed number of planes that are either

moving or stationary in a predictable pattern. Space-time manifolds are displaced by

varying degrees of mass and one can visualize a -ball on a trampoline where the

trampoline is displaced by the mass of the -ball   .Only the trampoline is on every

surface of the ball and the ball is curved by everything in the region of the

trampoline while the trampoline is curving or distorting everything around the ball.

Gravity is the curvature of space-time caused by mass. Gravity is conformal with

space-time and the conformations are determined by the mass exerted on space-

time exerted through gravitons and fermions (although spin 2 vector bosons are

Rked on the infinite dimension Hamilton-Jacobi Equations with regard to

viscosity solutions in an article 11.

also considered relating to gravity or the curvature of space-time.) Every point in

curved space with reference to time as the fourth dimension is attracted to every

other point as dictated by the mass of the point with reference to all other points   .

Also inertia is the resistance to push (or pull)by the mass ,so every point resists

push or pull as per the inertial (or resting)mass. This can be described by the Ricci

Tensor   used as a description for resting mass.

Space-time can be considered a metric or g as it acts with a vector and scalar

component as conformal to the relationship or inertial mass acting upon it as gravity

R g ab for the inertial mass R ab. The definition of a manifold based on surfaces

added to a three dimensional surface has ,in the past been related or restricted to

three or greater dimensions .However ,with the advent of flat matter and, elliptical

Josephson vorticies, noted as one or two dimensional if dilated time is considered

and is congruous to any matter that is superheated ,then the definition of a manifold

must be changed to include all dimensions of one or greater ,the first dimension

describes time and the second describes a closed flat string. Einstein predicted that

space-time is flat in the absence of mass and the geodesics of a space-time metric g

Rked on the infinite dimension Hamilton-Jacobi Equations with regard to   4
viscosity solutions in an article 11.

ab has been described as the Lorentzian metric in which it interacts as another

metric. When the expression of inertial mass(push)and gravitational curvature of

space-time caused by the inertial mass where the mass is described as a metric on

the remainder of the space-time manifold are equal ,the scalar of space-time

manifolds is zero and the manifold is flat(zero curvature).

It has been determined by this author that Space-time is directly proportional to

space and inversely proportional to mass.The dimension of perceived time slows

down as the object measured approaches a heavy mass.At the event horizon of any

black hole space-time shrinks towards zero without actually reaching it as an

asymptotic limit. Also space-time is directly proportional to space as in this universe

or space-time manifold time's arrow points forward not backwards .Therefore

space-time=space/mass x constant and the constant is 1/c2 as space/mc2 where

mc2 is the Grand Unification Energy of 10 19 Giga electronvolts is the equation of

"The Big Bang" at Planck Time(10-43 seconds).when space-time is -1/2 e –in

cotangent theta where n=number of dimensions and the dimensions approach

Rked on the infinite dimension Hamilton-Jacobi Equations with regard to
viscosity solutions in an article 11.

infinity as an asymptotic function ;the expression for space-time approaches zero

without reaching it as in the event horizon of a black hole. This is the quantum

ground state in a Relativistic Universe and unifies Quantum Mechanics and

Relativity.

CHAPTER ONE

Rked on the infinite dimension Hamilton-Jacobi Equations with regard to
viscosity solutions in an article 11.

BACKGROUND AND STRING THEORY

Strings are the smallest units postulated at Planck Length which is 10-33cm.

These strings are basically two dimensional units of either energy or matter

depending on which source you read and move in multiple planes. Strings can twist

turn ,rotate and connect or disconnect with other strings. It is postulated that are

made of two types ;open and closed. They move in 26 dimensions which are

compactified (rolled up and curled)into either 10 or 11 depending on whether the

reader incorporates Supergravity in Quantum Field Theory incorporating gauge

symmetry groups as illustrating the 11th dimension .Space-time is described in units

called orbifolds; which are manifolds or surfaces which twist and turn in the

configuration of a modified cone. A closed string represents a graviton particle or

gravitational movements   which are mimicked or represented by a spin 2 boson.

The world or universe   can be represented as a two dimensional sheet(The World

Sheet)of either closed or open flat strings that vibrate and rotate with reference to

themselves and each other in different combinations which represent twisted or

torsed donuts or toruses in flat combinations which can project into multiple planes

Rked on the infinite dimension Hamilton-Jacobi Equations with regard to
viscosity solutions in an article 11.

forming six dimensional twisted or puckered Calabi Yau manifolds. The two

dimensional world sheet mapped topologically with conformal mapping must apply

the rules of symmetry as well as quantum vibration and have to be dealt with even

in a relative vacuum even if this vibration self-annihilated instantly.

Because of the inability to empirically measure string activity   ,some of the

physics community looked askance at string theorists. Despite this   ,physicists such

as Brian Green are trying to quantify string theory with respect to 'The Big Bang' as

it was initially broached by Edwin Hubble or 'The Inflation Theory 'as postulated by

Alan Guth. Both theories were ex nihilo or "out of nothing" and were postulated as

occurring at approximately 13.7 billion years ago although one must differentiate

between Hubble Time and Conformal Time as differentiated by Einstein. Both

theories were tied in with the expanding universe as discovered by Edwin Hubble in

the late 1920's.

String Theory was originally purported in the 1980's when it was discovered

mathematically that nature follows harmonics of musical notes. These harmonics

were incorporated into two dimensional energy components of matter called

Rked on the infinite dimension Hamilton-Jacobi Equations with regard to   8
viscosity solutions in an article 11.

strings. Closed strings are continuous and formed loops ,double tori, torus

configurations(as mentioned previously),triangles, rectangles and a myriad of other

configurations based on the energy of the string with regard to other strings in

space-time. As mentioned previously the unit of space-time or geodesic is called an

orbifold which was defined earlier. Again this unit of space-time is a manifold or

surface that twists and turns in myriad configurations as do closed and open strings

.Calabi Yau manifolds are derived configurations of orbifolds and are generally

Planck Length or 10-33 cm. Open strings were coined by Dr .Roger Penrose as

twisters another description of closed and open strings. Twister theory incorporates

imaginary numbers with dimensions to indicate matter in a shadow universe

,similar but mathematically different from this's author's -1/2e-incotangent theta

incorporated into space-time with regards to quantum mechanics. At Planck Length

space-time is curved in and around itself like mini-black holes giving it infinite

curvature and transforming it into quantum foam from continuous space-time

according to quantum theory while Einstein purported that space-time was

continuous down to an infinitely small size. Also ,Einstein stated that what was what

Rked on the infinite dimension Hamilton-Jacobi Equations with regard to
viscosity solutions in an article 11.

appeared to be the force of gravity was in fact space-time curvature caused by mass.

This is why R g ab is the space-time curvature metric which describes the effect of

graviton space-time R causing curvature . String frequencies are based on the note

frequency f=1/2L(T/m)1/2 where L is the length of the string(guitar string not

string which is smallest unit of matter) (T is the tension and m is the mass of the

string .Decrease the length of a guitar string and the frequency increases .Place a

finger at the midpoint of the string and the frequency splits in two. These concepts

are the basis of string theory's correlations with harmonics or as Pythagorus stated

"the music of the spheres" which was first thought of by the Greeks. Open strings

split and join and closed strings mimic the spin 2 vector bosons(as previously

mentioned)which are quantized units of gravity or mass acting on curving space-

time geodesics .The spin 2 vector boson curve space-time as the carrier particles for

each and every target mass with the Ricci Tensor R ab. describing the resting mass

of the metric g ab. Again this is the effect of gravity rather than it being a force

according to Einstein .Hadrons were explained as strongly interacting elementary

particles. These spin 2 vector bosons appear as closed loop strings and as a basic

Rked on the infinite dimension Hamilton-Jacobi Equations with regard to
viscosity solutions in an article 11.

building block of matter spin 2 vector boson pervades all mass everywhere and

curve space-time as that mass Dimensionality of the space in which strings vibrate

relate to1-(D-2)/24which will only follow relativistic covariance if string moves in a

net of 0 dimensional space where a total of 26 non-compactified dimensions exist in

Quantum Field Theory and with the SO(32)gauge symmetry group this problem was

cleared up partially.C.Everett Peat p.99ClosedStrings appeared in type II string

theory like massless bosons with a spin of 2.This relates to the spin2 vector boson

involving the effect of gravity. Are strings flat matter or energy? The answer is

probably both.One could refer to the Higgs Boson which was coined "The God

Particle" which would bear out the boson as being the fundamental building block of

matter and energy all pervading .As previously mentioned there are five disparate

string theories which have duality to each other .They are type I ,type II ,type IIa,

Heterotic 8x8 and the SO(32) string theories which when observed all as one unit

forms Superstrings are groupoids of strings with the mathematical equations that

describes the way a sting moves and vibrates ,making sure they follow the rules of

relativity ,quantizing the equations of the relativistic string ,and making sure that all

Rked on the infinite dimension Hamilton-Jacobi Equations with regard to

viscosity solutions in an article 11.

strings are supersymmetric and follow the rules of gauge symmetry with strings

associated with gauge symmetry groups involving elementary particles .Quantum

Field Theory involves covariant systems where an action can move from one

coordinate system to another as in parity involved with the CPT Theorem which

involves covariance. The tension of a string was determined as 10^39tons Peat

p.101 making it appear to be more energy than matter with a miniscule mass

Although the m^2 operator for

$2\pi T$ for the open and closed strings determine it has a mass. $m^2 = 2\pi T \Sigma n =$

$1\infty\Sigma = i = 1\ to\ D - 2\alpha - n^{1\alpha n^{1}w} here 2\pi T =$

$\frac{1}{Regge}$ Slope which is $\alpha^{1.2\pi T}$ is a hige number but it's reciprocal is a small number..

Supersymmetry involves properties of iso-spin and strangeness of subatomic

particles . M Theory which is short for Matrix Theory or Membrane Theory

depending on the source Heterotic strings share two dimensions in one;one rotating

clockwise the other counterclockwise with regard to a rotating orbifold or Calabi

Yau Manifold..M theory contains supergravity in the 11th dimension(D=11)n the low

Rked on the infinite dimension Hamilton-Jacobi Equations with regard to
viscosity solutions in an article 11.

energy limit and reduces to the type IIa string theory which when compactified

forms a sphere.The infinite momentum limit of the D-0 brane may relate to the

U(N)super Yang Mills TheoryThe D-0 brane or membrane relates to the 10 which

broke up or cleaved into a dimensional state which has a limit of N approaching

infinity for 10 dimensional 0-branes.Based on this argument there were an infinite

number of dimensions in the vacuum pre Big Bang state and the 0-branes were the

building blocks of everything which were cleaved or broken into a six dimensional

component and a four dimensional component where the former is the Ca;abi Yau

Manifold and the latter is four dimensional space-time leading to the 10

compactified dimensions of Type IIa string theory. The idea is consistant with the

Law of Conservation of Dimensions which states that the sum total number of

dimensions of a system in any form is a constant. Infinite momentum(p) relates to

the Unified Energy with super gravity in the 11th dimension as described in

YangMIlls U(N).To be compactified into a circle or sphere type IIa string theory

would indicate world sheet which was two dimensional but spherical or circular in

two dimensions. Closed strings would fit better in a world sheet which was

Rked on the infinite dimension Hamilton-Jacobi Equations with regard to

viscosity solutions in an article 11.

compactified to a circle and this would eliminate duality as the five string theories

would curl up the world sheet to a point with infinite space-time curvature or a

circle of quantum foam when the environment is below Planck Length or 10-33 cm.

CHAPTER TWO

WHAT IS SPACETIME?

Space –time is a term that Albert Einstein coined for   what he considered four

dimensions of length   ,width, height, and time. It was based on the Line Element

ds2=dx2+dy2+dz2-c2dt2+dr2 where r=space-time curvature metric described by

the tensor R g ab. Ds2 is a description of the four coordinate system with regard to

space-time curvature and the relativistic effect of c2dt2. R g ab or r is determined by

R ab or the Ricci Tensor describing inertial mass of an object doing the curving and

the curving is done by the spin 2vector bosons and possibly gravitons or fermions .

Spiral space-time has a k=-I to the n cotangent theta power as suggested by Sir

Roger Penrose and proposed by this author. Flat space-time is space-time without

any curvature and occurs in a vacuum state. Considering the fact that photons have

a measurable mass while in a moving state moving photons can also curve space-

Rked on the infinite dimension Hamilton-Jacobi Equations with regard to   14
viscosity solutions in an article 11.

time as energy c ,however any mass at Planck Length or greater will curve space-

time. It is unknown if mass under Planck Length will curve space-time as there may

be a lower limit of Planck Length for space-time to appear in anything instead of

quantum foam or quantum dots which may be uniform or curved inward or

outward depending on the activity of string components. This quantum foam would

make up the orbifold unit of space-time.

Any mass curves space-time from string sized to that supermassive black hole which

centers each and everyone of the 750 billion galaxies in this universe and can be

displaced by the spiral or cone shaped designation of space-time as it approaches

the event horizon of a black hole where perceived time appears to the intelligent

observer to shrink along with local space to a point which may also be string sized

or 10-33 cm at the center of the black hole although there is no proof that the center

of a black hole isn't greater in size than that of a string.

Einstein's equation of Relativistic Gravity described the Einstein Tensor G ab=R ab-

1/2R g ab=8(pi)T ab where the Einstein Tensor=Ricci Tensor-1/2(Relativistic

Rked on the infinite dimension Hamilton-Jacobi Equations with regard to   15
viscosity solutions in an article 11.

Gravity)=8(pi)T where T=stress energy tensor of the metric g ab. Relativistic Gravity

is the curvature of space-time caused by the metric g ab and the Ricci Tensor

represents the inertial mass of the metric g ab. Inertia-gravity =1/2 but with the

metric tensors being abelian and anti-symmetric Inertia-gravity and gravity-inertia

are equal but opposite in magnitude and direction netting out zero with is the value

of the Einstein Tensor revealing a stress energy tensor of approximately zero.

The equation was derived by the Lorenzian Transformatios but basically says stress

energy is zero in the "Pre-Big Bang "or "Post Big Crunch "epoch .In black holes the

gravitational effect from a collapsed neutron star or galaxy is so great even light

can't escape. Beyond the event horizon of a black hole space-time is collapsed in a

spiral or vortex configuration to almost zero(as mentioned previously)and mass is

collapsed to almost infinite density. This mimicks the pre-Big Bang where the

Einstein Tensor G ab approaches zero.This is illustrated in the tensor expression of

The Equation of Everything where R abcd=R abc-1/2 R g ab/R ab where space-time

is curved inward by the collapsed mass of a black hole.In essence,the space-time

curvature metric is gravity acting on R,which is the spa ce-time that is curved by the

Rked on the infinite dimension Hamilton-Jacobi Equations with regard to   16
viscosity solutions in an article 11.

metric g ab. In the "Pre-Big Bang" epoch(up to 10-43 seconds or Planck Time)the

space-time curvature metric approaches infinite curvature as a point which has

infinite curvature and the extreme mass of the pre-Big Bang quantum bubble curves

extremely small space-time toward infinity without reaching it.In this case gravity

approaches half of a very large non-infinite number and anti-gravity from

antimatter approaches a very large non-infinite number but inertia equals the sum

total of the gravitational and antigravitational metric acting on space-time with

almost infinite curvature. This all indicates that Gravity isn't a force but the effect of

space-time curvature caused by any mass. The action of any metric in s[ace-time R

with the curvature metric of (-g)1/2. R can be described in terms of dimensions

such as d to the nth dimensional power and is extremely useful in string theory with

its postulated 26 dimensions compactified(curled up)to 10.As previously mentioned

Relativistic Gravity is described by 8(pi)Tab=R ab-1/2R g ab where R g ab describes

Relativistic Gravity and its effect on space-time. The space-time curvature metric is

described in the tensor equation R a b c d=R a b c -1/2r g a b/R a b where R a b c d

describes curved Lorenzian space-time R a b c describes flat Mintkowski or

Rked on the infinite dimension Hamilton-Jacobi Equations with regard to

viscosity solutions in an article 11.

Riemannian space and R g ab describes the space-time curvature metric acting upon

R a b c causing the curvature caused by the mass described in the Ricci Tensor R ab

describing the inertial mass of metric g ab.The entire expression is multiplied by

1/c2 to give the Relativistic form of the Equation of Everything .Utilizing Einstein's

equation of Relativistic Gravity and this author's equation of everything and solving

the stress energy tensor T ab one gets the gravitational constant G6.67x10-11

newtonmeters/sec2.Setting R ab from both equations equal to each other by the

transitivity postulate a=b b=c therefore a=c   Tij==T ab.T ba where i=initial event

and j=final event and mass=energy/c2 as T i j=

 T ab. Tba=Tij/||c2||2 and 8 pi T=R ab-1/2 R g ab via Einstein's formula where
8(pi)T ab.T ba/c4 reflects the stress energy tensor T ij=R ab-1/2R g ab at zero stress
energy.In other words the Einstein Tensor G ab=T ab.T ba=G(the gravitational
constant)as 8(pi)G/c4 is the gravitational coupling constant k and the stress energy
of zero has constraints of the pre-Big Bang or post-Big Crunch and at event horizons
of black holes. Stress energy is approached by the gravitational constant G=6.67x10-
11 newton meters/sec2 which approaches zero at the point of the Pre Big Bang"
epoch or where the Einstein Tensor approaches zero.The difference is attributed to
weak perturbations as described by the action formula S or the Hamiltonian
Operator for n eigenstates of energy as the 0 or null eigenstate is approached .An
operator is a complex function which operates on another function such as the
LaPlacean Operator ,which acts as a second degree differential equation of the
function it is operating onto the upward limit of the operator. An eigen-state is a
state of matter or energy with regard to the Operator that operates on another
function such as the quantum level of matter with regard to energy.
CHAPTER THREE
THE BIG BANG ;WHAT WAS IT?

Rked on the infinite dimension Hamilton-Jacobi Equations with regard to
viscosity solutions in an article 11.

OVER 13.7 BILLION YEARS AGO a quantum bubble with a 50:50 mix of matter and antimatter underwent either a 360 degree orb blast ,inflation ,or a two phase swirl which continuously expanded with progressive decrease in rotation. Prior to that is up to speculation. There is one theory that two membranes from other universes collided or touched triggering "The Big Bang' or a series of zero-BRANES the building blocks of matter underwent a dimensional recombination like "the popped bedsheet"hypothesis of Dr.Michio Kaku 1 forming a six dimensional manifold in a Calabi Yau configuration and the macroscopic four dimensional manifold of space-time. Another hypothesis is that a previous universe underwent a "Big Crunch" causing time's arrow to reverse and turning positive time into negative time(backwards)due to the implosion until it reached just before time zero when time advanced to Planck Time 10-43 seconds and the matter-antimatter mix exploded. The gravitational effect of antiparticles and particles may have caused "The Big Bang"and resulted in Dark Energy which propelled the accelerated expansion of galaxies away from each other that Edwin Hubble discovered in the late 1920's. According to the math antiparticles repel each other causing such an Rked on the infinite dimension Hamilton-Jacobi Equations with regard to viscosity solutions in an article 11.

explosive force in just under 50 percent of the matter antimatter mix that from a

Planck Length string sized quantum bubble the"Big Bang" results in a massive

antigravity surge from antiparticle antiparticle repulsion . This phenomenon not

only explains the existence of Dark Energy with pushes galaxies apart from each

other but also the paucity and lack of cohesion of antiparticles in the universe.

Antimatter was first discovered in 1932 and since then myriad antiparticles have

been discovered. Antiparticles are being isolated in Cern,Switzerland in the Hadron

Collider but as of this date the mass of any. antiparticle has not been conclusively

discovered. It has been noted that antiparticles such as the positron have the

opposite charge as matter particles but it has not been determined that antiparticles

have positive gravity. There are some experiments that do show repulsion of

antiparticles in positive gravity fields 2,but to collide particles and antiparticles can

cause annihilation of the particles with the expulsion of energy but this doesn't

necessarily mean that particles and antiparticles attract each other by gravity;only

possibly by charge and electromagnetism which are stronger forces than gravity

Rked on the infinite dimension Hamilton-Jacobi Equations with regard to   20
viscosity solutions in an article 11.

which acts as a weak force but isn't actually a force as previously mentioned. As

antiparticles self repel the energy required to push antiparticles toward each other

would be considerable and might destroy the antiparticles. If antiparticles mutually

repel with antigravity instead of attract with gravity a plausible mechanism for "The

Big Bang"can be made from the quantum bubble. At Planck Time 10-43 seconds a

50:50 mix of matter and antimatter caused an orb blast with a trajectory of theta in

the spacetime equation -1/2e to the i n cotangent theta power where i=the square

root of -1. And no is the number of dimensions either 4 for macroscopic spacetime

or 10 compactified dimensions including Calabi Yau Manifolds or Oribfolds in string

theory. The 2 pi radian orb blast with the mutually repulsive force of antiparticles

forcing an explosion converting over 99.999 per cent of the antimatter into Dark

Energy by the formula E=mc2 postulated by Albert Einstein where m=mass of the

antimatter. Whether the mass is positive or negative is up to speculation but the

Law of Conservation of Energy was purportedly been violated by the "Big

Bang"jnless the potential energy of the quantum bubble equaled the kinetic

energy,heat,and Dark Energy after "The Big Bang ".However the potential energy of

Rked on the infinite dimension Hamilton-Jacobi Equations with regard to
viscosity solutions in an article 11.

the quantum bubble didn't spontaneously appear "ex nihilo" or out of nothing and

may have been from a collision of a matter universe ,antimatter universe(or

membrane) and a relative vacuum forcing the pre-Big Bang implosion which drove

time's arrow backwards and reducing entropy from two universes to a quantum

bubble(making the occurrence a singularity as the Second Law of Thermodynamics

and Time's Arrow pointing forward must be suspended for this acr)As antiparticles

repel dark energy pushed outward in all directions carrying the balance of matter

with which it is subsequently attracted to other matter by gravity but not to

antimatter(which appears to have a lack of cohesion);the formula R j I k l-Rjl

kl(where kl is a superscript to the R in the second term while in the first R j I k l are

subheadings all to indicate covariant and contravariant tensors respectively=-

8(pi){Ge ji}}= g ji in four dimensions of space-time(an opposite curvature or

reciprocal curvature to gravity being antigravity as the sign for—(8 pi{[Ge j i]}=g j

iis negative.The rightside of the equation is the mutually repulsive force of

antiparticles and g j I is the metric of the antiparticle where i=initial event and

j=final event and-8(pi)G where G=the gravitational constant is from the Cosmologic

Rked on the infinite dimension Hamilton-Jacobi Equations with regard to
viscosity solutions in an article 11.

Constant ^which reflects the mutually repulsive force pushing galaxies apart

purported due to Dark Energy.E=2.71828 and e j I is the vector product of e from j to

i. This assumes a 360 degree or 2 Pi radian orb blast in the "Big Bang" and R is the

anti-gravitational effect on particles while conversely R j i is the antigravitational

effect on antiparticles. In a 360 degree or two pi radian orb blast the angle of

trajectory is pi radians or 180 degrees. The cosine of pi radians =-1 which explains

the -8(pi)G on the right side of the equation. There are approaching an infinite

number of 180 degree slices in a perfect sphere so the angle of trajectory for an

isotropic universe must be 180 degrees or pi radians.

THE C.P.T. THEOREM   AND ITS' INNATE SYMMETRY OF NATURE

Charge, parity and time whether reversed or not have innate symmetry in

nonlocal systems according to Quantum Field Theory. This indicates that if charge

were reversed as in the positron vs.the electron the magnitude of the charge would

be essentially unchanged. If time were reversed that

$$\Im\Psi(x, t)\Im - 1 = e\, iT\Psi(x, -t)\, where\ i =$$

Rked on the infinite dimension Hamilton-Jacobi Equations with regard to

viscosity solutions in an article 11.

$\sqrt{} -1$ and e is to the iΦ power, the transformation of t to $-$

t can be shown as commutative as the operator $\Im$ is antiunitary. The operator cancommute wit

The Hamiltonian and still reverse the sign of t.3 With parity X can be replaced with –
X and still have a commutative Hamiltonian Operator such that $(CPT)\mathcal{H}(CPT) - 1 = \mathcal{H}(-x)$. *Consider space* –
time curvature of matter and antimatter. Combining matter and antimatter have space –
time curvatures which would complement each other cancelling each other out resulting in fla

Space-time when matter and antimatter annihilate each other. This causes

energy=mass of the antiparticle+mass of particleXc2 resulting in the interfitting of

the reciprocal curvatures of space-time for identical particles and antiparticles

resulting in asymptotic flatness. In an antimatter universe of manifold pre-

dominantly anti-matter anti-gravity would attract rather than repel repel anti-

particles due to Parity replacing X by –X with the commutative Hamiltonian

Operator and in this case gravity would repel rather than attract particles for the

same reason. Conversely in a matter dominated universe anti-matter would repel

anti-particles with anti-gravity and matter would attract particles with gravity.

Rked on the infinite dimension Hamilton-Jacobi Equations with regard to
viscosity solutions in an article 11.

Outside of weak perturbations the symmetry of nonlocal systems is upheld with the

C.P.T. Theorem.

CHAPTER FOUR

TIME'S ARROW AND THE LAW OF ENTROPY

Time's Arrow states that time will be move forward in our space=time

continuum. The Law of Entropy or the Second Law of Thermodynamics states that

every system or subsystem will always go from a more ordered state to a less

ordered state. This obviously occurs in an open flat expanding universe ,but what

happens in the event that there is a "Big Crunch"? In an implosion   where a less

ordered state goes toward a quantum bubble where space-time develops a rip or

tear space=time can "pop" like a balloon and the second law of thermodynamics

may be violated. Mathematically,   the equation space-time=space/mass(1/c2)

negative space-time equaling negative space/positive mass x 1/c2 which dictates

the rate of compaction of space-time in a "Big Crunch". Also according to Einstein's

Law of Relativistic Gravity G a b=R a b-1/2R g ab where G a b=0 in a pre-Big Bang

or post Big Crunch epoch. Here R a b stays positive for the Ricci Tensor or inertial

Rked on the infinite dimension Hamilton-Jacobi Equations with regard to   25
viscosity solutions in an article 11.

mass but the space-time curvature metric known as gravity changes sign or the

direction of the vectors of the positive mass from –R g ab to +R g ab and the stress

energy tensor 8(pi)T a b goes equal in magnitude but opposite in direction .Due to

the fact that space-time=space-time and with the Bianchi Identity -1/2R g a b=1/2 R

g ab in magnitude but opposite in direction but since anti-symmetric they net out to

zero so the inertial mass of this universe stays the same as the Ricci Tensor R ab.

Since the only way to prove these conclusions is in a Big Crunch which could occur

in an accelerating rate toward Planck Time 10-43 or slowly where the only proof

would be a shift toward the ultra-violet with regard to the decelerating expansion of

galaxies as measured on the Hubble telescope. As approaching a heavy mass time

slows down the mass/per space decreases as in a Black Hole time must slow down

in a Big Crunch toward a quantum bubble and in that case "Time's Arrow" would

decrease. The equation $\mathbb{R}$ a b c d=R a b c -1/2R g ab/R a b changes to $\mathbb{R}$a b c d=R a b

c+1/2R g ab/R a b in the event of a Big Crunch as space-time undergoes reciprocal

curvature from the space-time curvature metric called gravity in other word space-

time would curve inward in the presence of inertial mass instead of outward and

Rked on the infinite dimension Hamilton-Jacobi Equations with regard to
viscosity solutions in an article 11.

with space-time reducing in size to a point or quantum bubble such as is analogous

to a black hole event horizon R a b c+1/2 R g a b is greater than R a b c. The entire

expression R a b c+1/2 R g a b is greater than R a b c d which is decreasing and as R

g ab=-R g ab as anti symmetric tensors then R a b c d=-1(R a b c-1/2R g ab/R a b)

whuch indicates the decreasing direction of time's arrow with the curved Lorenzian

or Riemannian Space-time $\mathbb{R}$ a b c d.

BLACK HOLE ENTROPY IS BASED ON STEVEN HAWKING'S FORMULA
S=2(pi)(NQ1Q5)1/2 where s= entropy N is the number of states or eigenstates in a
black hole postulated as 252 separate states and Q1and Q5 are the differential
charge between the first and fifth eigenstates.S=degree of disorder. The one brane
in M theory(membrane theory)is described with the monopole fixed negative
charge suggestive of an electron or positron with the antiparticle and the five-brane
represents 4 –space or curved Lorenzian Space-time which spiral into a black hole
event horizon .As a result the energy of a monopole acting on curved Lorenzian
Space-time across 252 states of matter reveal black hole entropy .Also S(black
hole)=Area/4 Length2 P=c3A/4Gh where h=Planck's Constant P=Planck Length of
10-33 mA=cross sectional area and G is the Gravitational Constant of 6.67 x10-11
newton meters/sec2 as c=3x10 8 meters /second or the speed of light which is a
true boundary for any mass as space-time shrinks to approach zero at that
boundary. Dr .Hawking postulated the entropy of a black hole to be 0.29 which
approaches zero. This indicate that as the cross sectional area approaches 0 or
P(Planck Length)the entropy of a black hole approaches zero(0)according to the
Bekenstein-Hawking Equation .This mimics the entropy(S) of the Quantum Bubble
at pre-Planck Time before "The Big Bang "where a 50:50%0 mix of matter and
antimatter are solidified by enormous pressure into what might be called a lattice
formation much as a diamond would occur. Indeed the central locus of a black hole
post event horizon ight have the same or similar configuration as with tremendous
pressures strange matter and liquid states of matter which would under other
ambient conditions not be liquid or solid. In a "Big Crunch" which may occur in
Planck's Time(10-43 sec)which is fast or a slow leak of space-time like a deflating
balloon would eventually approach 252 eigenstates of a quantum bubble similar to
that of a black hole and measured time would slow down as a shift toward the
ultraviolet would occur as galaxies recede instead of expand. A nuclear clock might

Rked on the infinite dimension Hamilton-Jacobi Equations with regard to   27
viscosity solutions in an article 11.

lose 10-3 seconds for each month that the galaxies recede instead of expanding and would be for most in perceptible except with scientific measurement. Also there would be no clear indication as to when the tail end of a "Big Crunch" would occur in the last 10-43 seconds where everything would shrink to approximately 10-33 cm like the initial quantum bubble. As a result it is extremely difficult to empirically prove that" Time's Arrow" is reversed in a "Big Crunch" also because the energy to reverse the sequencing of events in a universe would require the energy of a Big Crunch.

An alternative explanation for the end of this universe would be "Heat Death "where all matter and antimatter would slow down its' expansion until it eventually stops all fusion and fission in stars would eventually slow down and stop and everything would slow down to a crawl at 2.74 degrees kelvin(the temperature of the background microwave radiation from the "Big Bang")and all matter would "freeze".

CHAPTER FIVE;
WHAT IS SCHWARZCHILD SPACE-TIME AND HOW DOES IT RELATE TO BLACK HOLES?
It is well known that as the event horizon of a black hole is approached space-time approaches zero, time is dilated toward infinity(infinitely long)and mass increases dramatically along with gravity as space-time approaches a point string sized or 10-33cm in de Sitter or anti-de-sitter space. Quasars develop and spume out matter energy(Hawking Radiation)4 and information from the event horizon of a black hole whose origins or poles depend on the location, velocity and mass of the observer and are based on observational viewpoints.

The entropy of a black hole was already discussed in the previous chapter and again was postulated by Hawking as 0.29 where at least 252 different states of matter appear as N in the equation S=2(pi){NQ1Q5)1/2 where Q1 is the membrane boundary of the monopole(as mentioned previously)representing an electron cloud bounded by the membrane known as the 1-BRANE in terms of M Theory and Q5 represents Lorenzian Curved space-time or 4 space on which the 1-brane binds it with the electron(or positron)cloud acting upon it across the 252 states of matter.Q1 is 1.602x10-19 coulombs or the charge of an electron and a unit of space-time referred again as the oribifold and is generally bounded by Planck Length(10-33)cm. Therefore black hole entropy=2(pi)(252x10-19coulombs)(10-33 cm)to the one half power or 2(pi)(4.032x10-54 to the one half power or 2(pi)(4.032x10-108) Which approaches zero entropy. At the point at the event horizon the conformation is spherical and two dimensional with regard to position and action of the observer and the observer's motion. The mass of the information at the Event Horizon relates to its' spherical radius(which decreases toward zero)in Region II of collapsing matter using Schwarzchild Space-time. The event horizon is basically a cork which is a spherical region with extremely dense mass froma collapsed neutron star or galaxy and the hole of the event horizon is basically clogged up by the mass.In a way that matter queues up in that spherical region of collapsing space-time that approaches zero or string sized.Region 1 of Schwarzchild Space-time has a constant

Rked on the infinite dimension Hamilton-Jacobi Equations with regard to 28
viscosity solutions in an article 11.

radius and constant time.Region 2 has a radius approaching twice the mass and time approaching infinite dilation. Region 3 has the radius approaching twice the mass with time infinitely dilated in the negative direction .In other words time's arrow is reversed as in the tachyon .Region 4 has increasing time with regard to past and future time cones .The isotropic coordinates of Schwarzchild Space-time Metric is from the line element ds2=-(1-M/2r)2 divided by 1+m,/2r)2 and dt2+(1+m/2r)4{dr2+r2)d(omega)2 where omega is the Hubble Constant .Using the Kuskel Extension of Schwarzchild Space-time one has a symmetrical hourglass or cone shaped(spiral)configuration with the throat of the hyper surface at t=0 or infinitely dilated with the radius=twice the mass which is actually a 2-sphere or two dimensional hypersurface with one dimension suppressed. The topologic configuration of the hypersurface is RxS2 and is a circle shown with the radius equaling twice the mass. The surface above the throat at radius=2M lies in Region 1 and below the throat with r=2M in Region 4.There is a hyperbolic region proximal to the throat at the event horizon where r approaches 0 and time approaches infinite dilation.It has previously been determined by this author and the behavior of tachyons that negative mass goes backwards in time and reverses time's' arrow.Region 3 must be composed of negative mass as time's arrow is reversed as time approaches infinite dilationto the negative side while with ordinary matter(positive mass)it approaches time to positive infinity. SUBSTITIUTING a – MASS for positive mass in region 3 and reflecting it back to region 2 of Schwarzchild Space-time the t=infinity and t=-infinity cancel to time=0(zero)and the r=2mand r=-2m(minus 2 Mass)at zero space-time at time=0(zero),it can be concluded that mass and its energy equivalent cancel in regions 2 and 3 and only regions 1 and 4 are left.The information is not lost or destroyed but cancelled out with the negative mass cancelling the positive mass at the event horizon.The quasar effect is scattered throughout space from the 2 dimensional hypersurface over 2(pi)radians or 360 degrees of arc from a point in spacetime of zero entropy and dilated time over string sized space-time into curved Lorenzian Space-time with entropy being increased according to the second Law of Thermodynamics. The above is an explanation for "The Hawking Paradox"which states that the information absorbed by a black hole is lost(energy and mass)when the black hole eventually evaporates.

The negative mass must also produce antigravity and reciprocal curvature of space-time in region 3 which when superimposed on region 2 will push region 3 into region 2 in a form equal but opposite to region 2 resulting in asymptotic flattnessin regions 1 and 4. If one substitutes –mass for +mass in ds2(1-m/2r)2divided by (1+m/2r)2dt2 gets ds2=(1-m/2r)2divided by (1+m/2r)2times (1+m/2r)2divided by (1-m/2r)2 squared dt2.Taking the square root of both sides with negative mass substituting for positive mass with the Schwarzchild metric one gets ds2=1(dt)2or ds2=dt2 as the dx2+dy2=dz2 of the line element cancel out with the negative and positive masses at the event horizon with the negative and positive exteme masses at the event horizon from region 2 and region 3 being reflected upon region 2. This is consistent with space-time approaching zero at the event horizon and stops time at a point of Planck Length where the positive mass from region2 and negative mass of region 3 meet. Note in the cone approaching region 2 and region 3 from the opposite direction the radius gradually approaches zero and the

Rked on the infinite dimension Hamilton-Jacobi Equations with regard to viscosity solutions in an article 11.

mass and negative mass are sequestered proximal to the event horizon so the extreme progressive curvature of space-time to the infinite curvature of a string sized point is preserved. Therefore the net information going in and out of a black hole event horizon is a symmetric bi conar surface with region 3 cancelling region 2 leaving regions 1 and 4 to produce the quasar effect.(diagram eclosed)

CHAPTER SIX:

WHAT IS THE 'EQUATION OF EVERYTHING'?

There exists a simple mathematical relationship between space time and mass relating to gravity(space-time curvature metric) and the speed of light. In terms of a verbal description the relationship is simple. In terms of Relativity and Quantum Mechnics, the relationship is more complex.

As an object approaches an area of extreme mass such as a black hole, time slows down and eventually stops. As any object with mass approaches any other mass with is larger,time slows down even infanitesimilly. As a space-time with an atomic clock would approach the star(Solaris)or the sun, an internal clock would lose at least 1/10th of a second ,possibly more. And if possible to approach a black hole ,time would slow down through dilation towards zero where time would stop .Also space-time=space-time therefore space-time=space-space-time curvature metric(known as gravity) with inertial mass expressing the curvature. Therefore space-time is directly proportional to space. Also space-time is inversely proportional to mass as is proven by the action of space-time as it approaches the event horizon of a black hole. Therefore, space-time is directly proportional to space and inversely proportional to mass. The actual equation would be space-time=space/mass times a constant. This constant is $1/c2$ or 1/the speed of light squared as energy=mc2 and the denominator having mc2 becomes the Grand Unification Energy at the point of the "Big Bang" incorporating everything.

In terms of tensors R a b c d=R a b c-1/2 R gab/R a b where R a b c d is curved space-time or Lorenzian Space-time R a b c is flat Mintkowski or Riemannian Space-time R g ab is the space-time curvature metric known as gravity for the metric g ab whose inertial mass is described by R a b which is the Ricci Tensor of that mass.That ENTIRE EXPRESSION IS MULTIPLIED BY THE CONSTANT 1/c2 to give the equation of everything.

The Equation of Everything in terms of Relativity as postulated by Albert Einstein is that all motion is relative and not absolute. When mass(m)travels at approaching the speed of light boundary inertial mass approaches infinity ,space-time approaches zero(0) and length shortens to infinitely short or perhaps Planck Length according to the Lorenzian Transformations. As item A of mass m travels west in an environment that is traveling at velocity B when item A is traveling at velocity A the total velocity is the sum of A plus B.with vectors equaling the components of the motion away from 2 pi radians or 180 degrees such as (A plus B)cosine theta where

Rked on the infinite dimension Hamilton-Jacobi Equations with regard to viscosity solutions in an article 11.

theta is the angle in radians which is the difference between 2(pi)radians and the net angle displacement of A and B with regard to the surface or manifold .If there is another manifold or surface which is traveling at velocity C which if positive is added to velocities A and B cosine theta If C is negative or traveling less than velocity(magnitude and direction)A and B then velocity C is subtracted from velocities A and B cosine theta. If the manifold is moving in an expansion with a trajectory of 180 degrees as in the "post Big Bang" cos 180 degrees is one so the result would be velocity A and velocity B cos   theta +or – the velocity of space-time with regard to the stationary observer. The Equation of Everything is spacetime=space/mass times 1/c2 or $\mathbb{R}$a b c d+R a b c-1/2R g a b/R a b all times 1/c2. In terms of the metric g ab Lorenzian Curved Space-time or Riemann space-time is $\mathbb{R}$ a b c d as previously mentioned .R a b c describes flat space-time on which the metric of gravity R g ab curves space into curved space-time(as previously mentioned).The metric of gravity emanated from mass m whose inertia is described by R ab(Ricci Tensor)and this curves flat Mintkowski Space-time either inward or outward depending on the mass being acted upon by the metric of the mass doing the curving. The sum is described as R g ab where g a b is the metric of mass m.

The apace-time curvature metric of Einstein emanates from Einstein's Equation of Relativistic Gravity where a progressively increasing mass as described by the Ricci Tensor R a b-1/2 the gravity or space-time curvature metric which also increases with increasing mass as space-time curvature approaches infinity as in a point as in the space-time of the pre Big Bang quantum bubble(if below Planck Length)then a quantum foam described by zero-Branes (as previously mentioned).This depends on whether de Sitter Space has a hard boundary at Planck Length as in String Theory's Oribifold. The orbifold again is a twisted cone which is complex and can twist into a CalabiYau Manifold or surface which is a continuous surfacein a puckered appearance of a double torus that communicates with other Calabi Yau Manifolds 5 all six dimensional and in motion.

As increasing inertia and increasing gravity do not increase at the same rate as the speed of light boundary is approached space-time progressively curves to a point at v=c with infinite curvature. According to the Lorenzian Transformations infinite mass in dilated timeand reducing space-time with progressive increase in curvature causes inertia to increase at a greater rate than gravity because the metric g ab is acting on progressively increasing space-time curvature which is R g ab and the space-time curvature metric is half the inertia because the tensors are anti symmetric , abelian and space-time follows Bianchi's Identity .The covariant and contra-variant tensors of space-time are abelian with regard to the space-time curvature metric and as mentioned before are anti-symmetric as inertia approaches infinity at velocity approaches "c" without reaching it which is when the Einstein Tensor G ab=0 which is the stress energy tensor T ab at infinite space-time curvature. Anti-symmetric tensors cancel out in magnitude but with opposite direction.
THE EQUATION OF EVERYTHING IN TERMS OF QUANTUM MECHANICS HAS space-time=space/massx1/c2 described in terms of Planck Mass(the smallest unit mass for a quantum particle)or that two quanta can occupy and is approximately

Rked on the infinite dimension Hamilton-Jacobi Equations with regard to 31
viscosity solutions in an article 11.

1.22x10-24 kg .Mathematically Planck mass is the square root of hc/8(pi)G where h=Planck's constant at 6.63x10-34 and G is the Gravitational Constant of 6.67x10-11 newton-meters/sec2 and c=3x10 8 meters/sec and is the speed of light boundary. Space-time is described as the n-Dimensional state of a point-particle x at time t as a probability function and is operated on by the Hamiltonian Operator defined as – h/2mtimes the La Placean Operator with respect to the second derivative or d2/dx2+d/dy2+d2/dz2/ So utilizing the above space-time of H a}(r,t)|2 this is the probability density of a point particle "r" with respect to time or "t" in n-Dimensional Space. This equals -1/2e to the + or – I to the n cotangent theta power,here i=the square root of -1or -1e is the inverse or reciprocal of the natural log which is the integral of du/u which relates to the spiral fractal formula introduced in this authors first book "Mega physics ,A New Look at the Universe" and this defines the ground state in a Relativistic Universe as the natural log (l n 1)=0 and the natural log of infinity=infinity. Theta is the angle of trajectory at Planck Time from "The Big Bang" which is pi radians or 180 degrees. The infinity power of e(2.71828)is infinity and e to the negative infinity power is zero. Therefore e –in cotangent theta power defines the ground state as ln 1=0 and e I n cot theta is a reciprocal function and describes 1/0 which is infinity but e –in cot theta is zero where n=number of dimensions and theta is the angle of trajectory or pi radians.Therefore Planck's Mass(c2)H a|(r, t)|2 d n r where this expression is multiplied by the La Placean Operator in the n-dimensional state. A is the number of eigenstates of energy operated on by the Hamiltonian Operator (-h 2/2m)x La Placean Operator.In this case the Operator defines the wave function of r with respect to time(t) and also explains weak perturbations which explain quantum fluctuations in deSitter Space .Note also that the Hamiltonian Operator-minus h2/2mtimes the LaPlacean Operator)2+V o or initial velocity reflects momentum p=mv where p(rho) is Momenetum. So the momentum of quanta with Planck's Mass is incorporated over n-diemsnions from a to n eigenstates of energy and weak perturbations must include the momenta of quanta from the a=0 to a=n eigenstates of energy or energy levels. Therefore c2(hc/8 pi G)1/2+H a(eigenstates)|(r,t|)2 d n r times the LaPlacean Operator represented by the inverted delta to the n power equals =-1/2e +or –i n cotangent theta power with fluctuations which is again described by the Hamiltonian Operator in "a"eigenstates in the n dimensional state and ei n cot theta power is infinity. Therefore as 10 19 GEV approaches infinity.here n is the number of dimensions and I is the square root of -1. The Hamilton Operator in n dimensions describes the wave function of a point particle r with respect to time(t) in the n dimensional state with respect to "a" eigenstates of energy.c2(hc/8 pi G)1/2=10 19 power Giga Electron Volts which is the Grand Unification Energy which includes all forces except weak perturbations from quantum fluctuations with are corrected for by the Hamiltonian Operator. This occurs in the multiverse where the n-dimensional state approaches infinity and can be proven by subdividing a sphere to one second of arc or 1/3600 of a degree. This second of arc can be subdivided down to infinitiy and each subdivided portion is in motion as part of space-time with an infinite number of intersections of each unit or infinitely subdivided second of arc. As the intersection of two planes define a dimension and since the subdivided portions are not parallel due to space-time curvature when any

Rked on the infinite dimension Hamilton-Jacobi Equations with regard to viscosity solutions in an article 11.

even an infinitesimal amount of mass(non-vacuum)curves all space-time there are an infinite number of intersections from these subdivided lines or planes indicating an infinite number of dimensions in the multiverse.

When one second of arc is a plane in motion the topological surfaces mimic an osculating plane 7and each osculating plane from a topological standpoint define a dimension as space-time curvature causes an infinite number of intersections of these infinite osculating planes. According to Zeno's Paradox each degree is subdivided ad infinitum that prevent two solid objects from touching. With an infinite number of dimensions for space-time the left side of the quantum mechanics equation narrows from 10 19 giga electron volts toward infinity without ever reaching it as is true with the right side of the equation. There are also other theories such as M Theory which states the zero-branes which were building blocks to everything may have incorporated an infinite or near infinite number of dimensions .This situation applies for space-time to -1/2e –in cot theta and this applies to everything with it's reciprocal -1/2e i n ot cot theta power and this applies to zero space-time on the right for -1/2 e –I n cot theta. On the left side the Planck Mass is zero in the ground state as this is the vacuum or null state so c2(hc/G)1/2=0 and the zero-dimensional state states that the Hamiltonian Operator or point particle|(r,t)|2   d n r times the La Placean Operator in the zero dimensional state is also zero because the derivative of zero is zero. Therefore 0=0 in the ground state and the Equation of the Universe or Multiverse is upheld with respect to Quantum Mechanics. Note that this expression equals the Relativity Expression ℝ a b c d=R a b c-1/2R g a b/R a b times 1/c2 and due to the transitivity postulate that a=b,b=c therefore a=c indicates that the Quantum Mechanics Equation which is zero in the ground state and the Relativity statement which is zero in the ground state(as R a b c d=0 as Riemannian space nets zero when mass or inertial mass R a b approaches zero. Of course R a b c-1/2R g ab=0 ; R a b approaches zero forms the expression 0/0 which is everything as the case of curved Lorenzian Space-time. This is the case of where approaching the 0 dimensional case incorporates everything.

Kurt Godel   described a scientific tenet called "The Axiom of Incompleteness "stating that in any axiomatic system the set containing all elements must be incomplete. If there exists a set containing this subset ,the axiomatic system must be incomplete. Based on this tenet "An Equation of Everything "must be incomplete although it appears complete. An example of this is when 10 19 Gigaelectron volts(The Grand Unification Energy of all energies from "The Big Bang")can only approximate the value of infinity and while it is true the 10 19 Gev approximates infinity due to weak perturbations from quantum fluctuations it isn't definitive. However in the infinite dimensional case where 0=0 as 10 28 power in the denominator of the left side and infinity in the denominator of the right side approximate 0=0.This will again be brought into focus later. Quantum Mechanics and Relativity are unified with the equations ℝ a b c d=R a b c-1/2R g ab/R ab times 1/c2 and Planck Mass times the expectation value of the probability of a point particle r at time t in the n dimensional state where space-time is -1/2 e –i n cot theta with the reciprocal curvature being -1/2e i n cotangent theta where n=the

Rked on the infinite dimension Hamilton-Jacobi Equations with regard to viscosity solutions in an article 11.

number of dimensions netting infinity divided by infinity which is everything except zero which is the null state or absolute vacuum state indicating in this total case that nothing or spaceless ness doesn't exist. Based on measurements the integral of du/u mathematically indicates -1/2 e –i n cotangent theta power as space-time with the curvature metric R g ab while the sum total of both reciprocals for space-time would result in flat space-time as in a vacuum which is precluded by the expression infinity/infinity because it doesn't include zero and is therefore incomplete .The Quantum Mechanics equation of everything

$$\text{is}\frac{\dfrac{\psi(r,t)dn(power)r\nabla n}{c2\left(\frac{\hbar c}{8\pi G}\right)1}}{2} + \mathcal{H} \; from\; a\; to\; n\; eigenstates(|r,t)|\; d\; n(power)r\nabla n =$$

$$\psi(r,t)d\,(n\;power)r\nabla(n\;power) - \frac{1}{2}e - i\,n\cot\theta \quad equals\; the\; Relativity\; Equation\; R\; a$$

b c d=R a b c-1/2R g a b/R a b times 1/c2 and at the ground state they both equal zero and each other and can be simplified to space-time=space/mass all times k=1/c2 where c=3x10 8 meters/sec.

The Hamiltonian operator
$$\mathcal{H} =$$
$$\hbar\frac{2}{2m}\nabla\; to\; the\; nth\; power\; where\nabla\;(nabla)is\; the\; LaPlacean\; Operator\; handle\; the\; weak\; perturbation$$

or quantum fluctuations acted upon by the wave function of the point particle r at time t. This will be delved into again later in this book.

Another equation postulated was the wave function$\psi(r,t) = \int e \;\; \frac{i}{\hbar}$ to the integral
$$power\int \quad (\frac{R}{16\pi G} + \frac{1}{4F2} + \psi iD\psi - \lambda\varphi\psi\psi + D|\varphi|2 -$$
$$V(\varphi)where\; \psi is\; bar\; \psi\; or\; a\; probability\; function. This\; is\; based\; on\; the\; Yukana\; Equation, Relativit$$

The Schrodinger Equation and the book "Quantum Mechanics and Path Integrals by Richard P. Feynman and Albert R. Hibbs. The path
integral
$$\oint \quad \nabla 2\psi(r,t)applies\; Poisson'sEquation\; for\; the\; path\; of\; dual\; vector\; field\; 4\pi\rho\; where\; \rho\; is\; the\; en$$

Energy density of matter and
$$\psi(r,t)is\; the\; wave\; function\; of\; point\; particle\; r\; at\; time\; t forming\; a\; path\; integral of\; 16\pi\rho 2\psi(r,t)$$

$$16\pi\left\{\frac{\rho 3}{3}\right\} -$$
$$\frac{\rho 3}{3\psi(r,t)superimposed}with\; parallel\; transport\; to\; curved\; manifold\;(surface)with\; space -$$
$$time\; curvature\; of\; -\frac{1}{2}e - i\,n\; cotangent\theta\; for\frac{\rho 3}{3}and\; space -$$
$$time\; curvature\; of\; -\frac{1}{2}e +$$
$$i\,n\; cotangent\; theta\; for\; -\frac{\rho 3}{3}again\; approaching\; the\; ground\; state. The\; -\;|D\phi|2 -$$

Rked on the infinite dimension Hamilton-Jacobi Equations with regard to viscosity solutions in an article 11.

V(ϕ)relates to the Higgs Field which relates to the spin 2 vector boson which curves space — time by any mass as described by the Ricci Tensor.

CHAPTER SEVEN
WHAT IS M THEORY?

M Theory is a combination of the five extant string theories ;type I, Type II ,Type IIa ,Heterotic 8x8 and the SO(32) string theories. The five string theories are dual to each other as previously mentioned mathematically observing the same phenomenon from five different vantage points or approaches all describing the same thing. What makes M Theory the combined five string theories is the incorporation of membranes which vibrate. These membranes are submicroscopic and may be string sized 10-33 cm or possibly smaller. A membrane is almost continuous with any and all matter and possibly energy and are continuous across the different dimensions whether 26 compactified(curled up)to 10 as in type I string theory or an approaching infinite number of dimensions if the osculating plane can be applied to an infinite number of non-parallel planes which are subdivided from 1 second of arc(1/3600 degree)and in motion expansion and rotation as in Godel's Rotating Universe .These non-parallel planes must intersect an infinite number of times if they are non-parallel due to the space-time curvature metric forming an infinite number of dimensions in the osculating planes. Membranes would have to be contiguous with these osculating dimensional planes and would contain all matter including energy which is converted to matter as matter=energy/c2. M Theory was originally coined for Membrane Theory and Matrix Theory depending on the source read and the different membranes called in short hand branes describe different states of matter interacting with energy in space-time .M Theory has no clear cut definition except the duality with all five string theories although when utilizing super gravity for the existence of the 11th dimension instead of the compactified 10 dimensions in the low energy limit involving D-0-Branes it reduces to type II a string theory when compactified by
$2\pi R$ where R measures the limit to infinity of$D − 0 −$
Branes in the infinite momentum limit of MTheory as in the case of $U(N)$in the super Yang Mi

Theory 9. The compactified type IIa string theory is a sphere of Radius R. The D-0-Brane has a momentum limit of 1/R and a bound state of D-0_branes have a conditional momentum of N/R where N relates to the unified Yang Mills state relating to the gauge limit in terms of d4space such that g2YM N approaches infinity. Note that D-1-Branes or 1-branes describe the string as well as the monopole depending on the text read.M theory incorporates super gravity and the 11th dimension to string theory."branes "'are movable membranes that incorporate special dimensions and particles including strings. Membranes as mentioned previously vibrate move and are real and measurable .Charges ,state and tension are incorporated into membranes which extend up to at least the ten dimensions of

Rked on the infinite dimension Hamilton-Jacobi Equations with regard to   35
viscosity solutions in an article 11.

string theory. The limited definition of M Theory is "the limit of strongly coupled IIA string theory with 11 diemsnional supergravity or the Poincare invariance"The 2-brane or M 2-brane couples to the potential(V) of eleven dimensional supergravity .The 5-brane or M-5 brane carries an electromagnetic charge of the potenetial(V)of eleven dimensional supergravity.The 5-brane carries the potential which is coupled to the 2-brane. The Neveu-Schwarz 5-brane carries the electromagnetic charge(V)of what is known as the NS-NS 2-form potenetials where the NS boundary state is a fermionic field that is anti-periodic on the world sheet in the closed string or the double of the open string. The Neven-Schwarz algebra is a world sheet algebra with regard to the energy momentum and super-current tensor into a sector where supercurrent is antiperiodic and modes are half integer valued .Therse are what are known as Fourier modes .The null state is orthogonal to all physical states including its' own. Incorporating the anti-periodic fermionic field which is the NS boundary state onto the 4 and 5-brane representing space-time in the four macroscopic dimensions would have the Neven-Schwarz 5 brane incorporated with the NS-NS 2-form potentials on the world sheet(which represents flat or two dimensional matter) and is divided by the Ricci Tensor with regard to the sum total or infinite sum of all inertial mass to equal the Lorenzian tensor of space-time in terms of supergravity incorporating 11 dimensions. The momentum limit of the inertial mass as it approaches the infinite momentum limit where "R" approaches a very large number is incorporated into the denominator of space-time=space/mass .Space-time curvature based on the infinite momentum limit R where N approaches infinity in the unified Yang Mills state in the d4 state relates to the macroscopic gauge limit and the D-0-brane which is bound with conditional momentum of N/R. The momentum limit of 1/R of the D-0-branes relates to the infinite momentum limit and curved Lorenzian space-time or Riemannian space-time where the 256 permutations net out to zero in the ground state. Applying this to 11 or possibly more dimensions(infinite dimension state as previously mentioned in terms of M theory is more difficult.

SUPERGRAVITY which incorporates the 11th dimension into superstring or the IIa closed string theory to compose M Theory where the IIa closed string theory compactifies to a circle or sphere is involving global supersymmetry which produces a Yang Mills Gauge Group of Osp(1/4).The number of states within 11 dimensional supergravity are eAM implies ½(9)x(10)-1=44 ψM $implies$ $\frac{1}{2(9x32-32)} =$

128 $and\ A\ MNP$ $\binom{9}{3}$ =84.where M and N represent 11 dimensional curved space indicies with 32 dimensional spinors. The term e A M where A is the superscript and M is the subscript and it represents a linking or parallel transport of the base manifold or surface with the tangent
space.

ψM $is\ the\ graviton\ field\ and\ A\ MNP\ is\ an\ antisymmetric\ tensor\ field\ with$ 128 $boson\ fields\ eq$

equalling the fermionic field giving a total number of 256 boson and fermionic fields which is the number of fields in the 11 dimensional N=8 model for super gravity again where bosons and fermions curve space-time.

Rked on the infinite dimension Hamilton-Jacobi Equations with regard to viscosity solutions in an article 11.

As there are 256 bosonic and fermionic fields curving flat Mintkowski or Riemannian space-time which has 256 permutations and since Riemannian space-time graviton or bosonic fields are anti-symmetric tensor fields acting on flat space-time as anti symmetric tensor fields the net permutations of gravity space-time curvature on flat space net out to zero which is the numerator of space/mass=space-time .Regardless of the denominator even when multiplied by the speed of light squared the net result is zero which is curved Lorenzian or Riemannian Space-time.Space-time in the "null state "or equivalent to the Einstein Tensor G ab. Therefore "the Equation of Everything" does apply to 11 dimensions as incorporating supergravity. Note in the above Riemann Space-time can be applied as curved or flat when Mintkowski Space-time is always flat.Riemann postulated that all mass can be described as curves in space-time although flat space-time means that a missal shot out in flat space will theoretically never return to the starting point but with curved space-time it will eventually return to the starting point as in Einstein's Closed Curved Universe vs. Friedmann type II open flat expanding universe.

In terms of M-Theory the Neven-Schwarz 5- brane incorporated with the NS-NS 2 potentials incorporating the anti-periodic fermionic field(vacuum state of ferminon field) on the world sheet would go into the numerator of space/mass and the infinite momentum limit in the bound state of D-0-branes would go into the denominator suggesting the inertial mass times the speed of light squared of the quantum bubble with approximately 750 billion strings acting on D-0-branes in the pre Planck Time epoch of under 10-43 seconds. The world sheet would be 2 dimensional as are strings and the anti-periodic fermionic field density in terms of a tensor field and applying Poisson's Equation for dual vector fields   would be applied via parallel transport onto the flat manifold of the world sheet. Since the energy momentum and supercurrent tensor are also anti-periodic the energy momentum R goes into the denominator with the supercurrent incorporating into the 1-brane as a carrier such as the monopole or electron cloud while the fermionic field density is incorporated in the numerator as previously mentioned. This incorporates to enclose a near infinite momentum onto an accelerating expanding space-time acted upon by gravity (fermionic fields caused by the inertial mass of the quantum bubble).AS A RESULT N-BRANES SUGGESTIVE OF LORENZIAN SPACE-TIME=D-0-BRANES+ and –Neven -Schwarz 5- Brane incorporated with the NS-NS 2 potentials/N/R where R is the infinite momentum limit and N=1. The N-Brane as an infinite number of dimensions are approached but never reached approaches infinity but is never reached and since
anything/1/
∞ is infinity due to the infinite momentum limit so as applied to N dimensions or the N − brane infinity =
infinity with supergravity localizing the Einstein Tensor G ab which is 8πT which is the stress

Energy tensor of matter acting by the Poisson Equation and is zero(0)based on the D-0-branes.

Rked on the infinite dimension Hamilton-Jacobi Equations with regard to 37
viscosity solutions in an article 11.

CHAPTER EIGHT
WHY THERE IS AN ARGUMENT THAT THERE ARE AN INFINITE NUMBER OF
DIMENSIONS

   A dimension is formed by the intersection of two planes which have an
intersection of an angle which in length, width and height is 90 degrees or pi/2
radians. However other dimensions may have different angles of intersection which
aren't perpendicular .For example ,the intersection of the x ,y ,and z axis has a near
infinite number of points which while discontinuous can be subdivided down to an
infinite number of intersections and the limit of these subdivisions can blend or blur
into a continuous dimension. There are a postulated 26 dimensions associated with
string theory and superstring theory and these dimensions can be
compactified(rolled or curled up)to 10 dimensions or 11 dimensions with
supergravity included. To make a corollary in logic one must depend on axioms as
assumptions. These axioms must be assumed to be true and if false the corollary can
be false. For example if it's an axiom that planes are stationary and later it is found
that planes are in motion, then the corollary that there are three dimensions length,
width and height with no others as in Euclidian Geometry is false. In the case of
space-time this corollary is false as time is considered a fourth dimension by Albert
Einstein and his space-time curvature theory which was proven with the "solar
eclipse" experiment in 1921 to measure a change in position of an object near the
sun(which is of significant mass compared to empty space)proves that space-time is
curved and not flat(without curvature).It was mentioned previously by this author
that each subdivision of a second of arc(or $1/3600^{th}$ of a degree)can be subdivided
down toward an infinite number of cuts. If these cuts of space-time were totally
parallel to each other as if space-time were totally flat and without curvature, then
these cuts would never intersect each other. However because space-time has been
postulated as being in motion along with mass, expanding and rotating(according to
this author and Kurt Godel),then these subdivided planes would not be absolutely
parallel but would be almost parallel, where the parallel nature is distorted as space
approaches a reads of extreme mass such as a black hole where the curvature of
space-time increases towards infinity which is why space-time appears to spiral into
a black hole in a cone like configuration as the event horizon is approached. The
time component or fourth dimensions shrinks toward zero as the extreme mass is
approached as space-time is inversely proportional to mass.
   Now if there are a near infinite number of planes from the subdivisions of each
second of arc and is these planes are osculating and in motion expanding with a
progressive decrease in rotation from the area of maximum rotation at just before
Planck Time 10-43 seconds and approaches pure expansion with only a miniscule
rotation as 13.7 billion years later ;then these planes are osculating into expansion
and rotation vectors with the unit tangent vectors being measured along these
osculating planes. Obviously ,if all these axioms are true ,these near infinite(if there
is no downward boundary to space-time at below Planck Length or 10-33 cm)and if
the quantum foam can also be subdivided down to infinity then there would be an

Rked on the infinite dimension Hamilton-Jacobi Equations with regard to   38
viscosity solutions in an article 11.

almost infinite number of intersections between the near parallel planes formed by each subdivision from 1 second of arc downward toward infinity .As Einstein showed that these planes of space-time are curved and "The Big Bang" Theory shows them in motion ,these are osculating planes and the unit tangent vectors reflecting the curvature of these planes and the planes will intersect a near infinite number of times forming a near infinite number of dimensions. This was all explained earlier in this book and there are other arguments to this premise as well. This conclusion would be totally consistent with the infinite or near infinite universe theory and would preclude that D-0-branes are not at absolute zero dimensions but an infinitely small number approaching the zero dimensional state as it is when an object approaches 0 degrees kelvin but cannot be reached .Indeed the null state or D-0-branes are only an approximation where the zero dimensional state is approached as an asymptotic function as with limits where the size of each extant dimension is reduced toward zero without reaching zero. The reason why the axiom nothing doesn't exist when according to "The Big Bang" ex-nihilo states it does is because of these limits in size downward for the dimensions which are in motion with the osculating planes never reaching the zero dimensional state as when "time's arrow" reverses time from before the Big Bang when a Big Crunch would cause time to move backwards in a massive implosion bringing it back toward zero time without ever reaching it before the Big Bang moves time's arrow forward again. These subdivisions can be curled up or compactified toward zero BUT ARE NOT ZERO ONLY APPROACHING ZERO AS AN APPROXIMATION .Also the diverse motions of string and superstrings in space-time are diverse but the number of motions from a vibration to a twist,partial rotation(degrees can also be subdivided plus the motions of strings can be subdivided)make the two dimensional world sheet as an approximation as would the two dimensionality of flat matter or flat anti-matter .Indeed, the other dimensions are there but so miniscle and compactified that for the sake of math they aren't important and there limits can be brought to zero.

Infinite Dimension Symmetry involves the symmetry transformation of space-time as. generated by similarity transformations of T ab which is the stress energy tensor of theories associated with conformal gravity C.A type of algebra exists called infinite dimension subalgebra which is associated with Lie Groups as a subgroup of Lie Algebra. Infinite dimension subalgebra has nonsingular commutators and uses gauge symmetry of the Yang Mills groups to form a weighted tensor algebra. An approaching infinite set of conformal fields have well definied commutators of an arbitrary pair of zero modes constituting an infinite dimension subalgebra with regard to the full symmetry of string theory and M Theory(including the compactified IIa string theory which is spherical .Infinite dimensions do not arise through the moding of a finite number of conformal fields but rather than an infinite number of conformal fields of space-time acted upon by any mass.W

∞[17]*in terms of field equations with Wi*[18]*as conformal weight W∞ must retain all modes (time space −*

time while zero modes form only Cartesisan Subalgebra. This subalgebra possesses the proper

Ties of string symmetries and is a supersymmtery in that it's generators do not commute with the generators of the Lorenzian Transformations and comingled or quantum groups which are excited and with a different spin it is spontaneously broken in flat space-time because NOT ALL GENERATORS COMMUTE WITH THE STRESS ENERGY TENSOR(T ab)relating to the free scalar and it transforms excitation of differing masses into one another .Symmteries or gauge symmetries should be local symmetries and the constant tensors
ψ and χ in the generator were depending on the scalar field $X\mu(\sigma)$ where mu is a superscript th

Locality would be apparent,nut it is x dependence that affected commutations. Infinite dimension subalgebra would be a global part of Gauge Symmetry and must include both holomorphic and anti-holomorphic derivatives with propagating degrees of freedom generated by operators evenly balanced between two types of derivatives.Short distanc e singularities e ipxx L(x)ei qXL(w)equals the fraction e i(P+Q).xL(W) /(Z-w) to the-p .q power and e ip.x(=L(x) is also a power with e iqXl(w) as a power.The stress energy relates to (Z-w)to the –p.q acting upom e i(P+Q).Xl(W) which relates to e –in cotangent theta of space-time, The infinite dimension state where Napproches infinity forces the N-brane of M theory to approaches
an∞Brane for Curved Lorenzian Space $-$ time R a b c d and the superequation

{Ha to n eigenstates|(x,t)2d nr
$\nabla\, n$ divided by $c2\sqrt{}\ \hbar \dfrac{c}{8\pi G} =$
$\nabla n|x,t|2d\, n\, x\nabla x \left(-\dfrac{1}{2e} - \right.$
in cotan theta$\Big)$ where n dimensions approaches ∞ therefore $-\dfrac{1}{2}e -$
$i\, n\, \cot\theta$ approaches $e - \infty$power so the wave $\dfrac{function\psi|x,t|2d^{\frac{n(power\ of\ )x\nabla n}{-1}}}{2} e -$
$i\, n\, \cot thea = \psi|x,t|2dn\dfrac{x\nabla 2}{\infty} = 0.$ Here space $-$ time is described as $-\dfrac{1}{2}e -$
$i\, n\, \cot\theta$ where theta approaches π radians or 180 degree trajectory of a sphere describing

the Big Bang. Note the particular math of infinite dimension su-algebra gets extremely involved and can be referred to by the footnote 10.

Call{ } the set of all covariant tensors in d –dimensional space.The operatorΔr acts on the subset$\{\psi(k), w\, i\}$such that $\Sigma l = 1$ to $r\{\psi(h), w\, i + \delta il\}$where$\psi$ is the wave function of w i and here i is the initial event the element of algebra of

Rked on the infinite dimension Hamilton-Jacobi Equations with regard to viscosity solutions in an article 11.

Pairs such that$\{\psi(h), w\,i\}$ modulate relations $\rightarrow \lambda\{\psi h, wi\} + \omega\{\psi(h), w\,i\} - \{\lambda\psi(h) + \omega\psi(h), w\,i \rightarrow)$ such that $\Delta h\{\psi(h), w\,i\} \rightarrow 0.\psi$ is a subset of S. In this case h is not Planck's Comstant but a variable power to the wave function. S, η is the Mintkowski Metric for flat space $-$ time. $|S|$ are the elements of S; $S - \psi$ are the complement of ψ M.S. Conformal weights $\omega 5 - \psi + V(t) - P(u)$ were conformal weights of ψ that don'tcorrespond to the indicies in ψ together with the ı

Weights of
χ that don'tcorrespond to the indicies in ψ with the conformal weight of χ that don'tcorrespon

To indicies in the image of the wave function of w and P(U) such that $\omega s - \mu + V T = P(U)$ Wick'sTheorem states the sum over μ and P is a sum of all possible contractions of the te

Tensors
ψ and χ The conformal weights are those of the noncontracted indicies except the weight

Of
ψ and are increased incrementally by the weights of the contracted indicies through the powe

Of the operator
$\Delta | S - \mu |$ Thet wo point function for free bosons implying the curvature of Mintkowski space $-$ time by any mass is $< x\, m(z) x V(w) \geq -\log(Z - w)\, \delta v to\mu$ where δ is the superscript and μ is the subscript and the $-\log(z - w)$ is to this δ where $\mu$is the covariant tensor and$\delta$is the contravariant tensor all aplied to the ı

Boltzman Equation incorporating states of matter with regard to bosons and conformal gravity. Wick's Theorem states that short distance singularities as at the event horizon of a black hole form a conformal block or non-holomorphic operator or sum of the products of holomorphic and anti-holomorphic fields constructing a full commutator out of the commutators for anti-holomorphic components .Holomorphic and anti-holomorphic operators are constructed from mutually

Rked on the infinite dimension Hamilton-Jacobi Equations with regard to viscosity solutions in an article 11.

commuting sets of the creation and annihilation operators .Short distance
singularities at Black Hole Event Horizon are eip.xL(z) where ip and x L(Z) are
powers times e iqX L(w)=the quotient of e to the i(p+q).xL(w) power/(Z-w)-p.q .p.q
must be integers .Products of holomorphic and anti-holomorphic fields constructing
a full commutator out of commutators for anti-holomorphic components form the
conformal block on space -time|S| giving a mechanism in terms of math on how
space-time is crammed down by annihilation operator in terms of bosonic effect of
extreme gravity on asymptotically flat Mintkowski Space-time
s,η *as event horizon is a black is reached.* This may appear far a field from infinite
dimensions but the stress energy tensors relating to the annihilation and creation
operators acting on Mintkowski Space-time clearly reveal that space-time or the n-
brane =space(D-0-Branes + or – the Neven Schwarz 5-brane with NS-NS2
potentials/N?R where R is the Infinite Momentum potential as N approaches
1(which is mass or the Ricci tensor x c2).They also explain why space-time spirals
down to a point (string sized?)without actually disappearing. The Hamilton-Jacobi
Equation for infinite dimensions is such that $\lambda v(x) + < Ax = \phi(x), Dv(x) >$
$+ H\big(A\mathcal{B}power\ x, D\ v(x)\big) = 0\ where\ x \in X\ wher\ e\ X\ is\ a\ real\ Hilbert\ Space, \lambda >$
$0\ and\ H: H\ by\ X\ is\ continuous\ as\ a\ closed\ linear\ operation\ with\ a\ compact\ and\ dense\ inclusion\ I$

D(A) is less than X. We assume A is positive and self
adjoint.$\phi: D(A\ \beta\ power) implies\ D(A -$
$beta\ power) and\ is\ Lipschitz\ continuous.\ Piermarco\ Cannarsa\ and\ Maria\ Elisabetta\ Tessitore\ w$

CHAPTER EIGHT

The infinite dimension Hamilton-Jacobi Equation and applied specific situations is in

a paper written by Piermarco Cannarsa and Maria Elisabetta Tessitore called

"Infinite Dimensional Hamilton-Jacobi Equations and Dirichlet Boundary Control

Problems of Parabolic Type".

The stress energy tensor of a free scalar T a b$\longrightarrow T(\sigma) and\ T(\sigma') = \frac{1}{2}: \partial x \partial x(\sigma) :=$

$\lim \epsilon \longrightarrow \frac{0i}{2\partial x(\sigma)\partial x(\sigma+\epsilon)} + \frac{1}{4\pi\epsilon 2} where\ the\ commutator\ is\ 4[T(\sigma), T(\sigma'] = \lim\ \epsilon, \epsilon' \longrightarrow$

$0[\partial x(\sigma \partial x(\sigma + \epsilon)]\partial x(\sigma')\partial x(\sigma' + \epsilon')\}.\ for\ real\ numbers.\ or\ F(\sigma\delta(\sigma - \sigma') -$

$f'(\sigma')\delta(\sigma - \sigma') \longrightarrow$

$[T(\sigma), T(\sigma')]which\ applies\ to\ the\ commutation\ of\ the\ stress\ energy\ tensor\ where\ \sigma\ and\ \sigma' have\imath$

E stress   energy as related to T a b where g a b relates to sigma and sigma prime.

Incorporating imaginary numbers $2iT(\sigma')\partial'(\sigma - \sigma') - iT'(\sigma')\delta(\sigma - \sigma' - \epsilon')/(\epsilon +$

$\epsilon')2 + 2\delta(\sigma - \sigma') - \delta(\sigma + \epsilon - \sigma')/(\epsilon + \epsilon')3.$ Stress Energy between   sigma and

sigma prime yields epsilon relating to a small number for stress energy relating to black hole entropy(s) using Virasuro Algebra: e ip .$X(\sigma)$ :: e iq. $X(\sigma + \epsilon)power =$

: e ip dot $X(\sigma) + iq$ power dot $x(\sigma + \epsilon)$: e p −

$\frac{q}{2\pi}$ power. As ϵ is greater than zero the entropy of a black hole approaches zero by the normaliz

normalization coefficient with normal ordered operators .The stress energy

between sigma and sigma prime is a maximum at the event horizon or a black hole

where space-time approaches epsilon.11

As is well known
s=r(theta)
θ where s is   the distance or arc length subtended by a sphere r is the radius of the cut of the

If the sphere is compactified and converted to two dimensions with regard to the

world sheet as in type IIA string theory converts to a description in which M Theory

may be derived .As theta

θ approaches zero degree but not reaching it the cuts becomeinfinitely small. The radius rema

remains constant. As the sphere describes space-time in the Big Bang rotating and

expanding according the H the Hubble Expansion Coefficient the cuts "s" approach

zero initially then expand as theta approaches zero then expands and "r" describes the radius of this universe .If the near infinite number of cuts remained parallel

theta would remain very small (below one second of arc).The radii would approach

being parallel without reaching it as the angle between radii approach zero degrees

without reaching it. The cuts are not parallel due to the extreme mass of the

quantum bubble with extreme Dark Energy overcoming the extreme mass of the

bubble curving space-time to infinity as in a point which is a two dimensional

sphere composed of strings. The unwinding curvature of space-time will reach the

space-time curvature metric of Einstein instead of that of asymptotic flat space.

These infinitely small cuts with ds approaching zero as theta approaches zero with r

or the radius continuously increasing according to the Hubble Expansion Factor

forms an infinite number of dimensions as the curvature causes an infinite number

of intersections of these non-parallel infinite number of planes formed by the

infinite number of cuts in the circle (2D) or sphere (3D) or space-time (4D) and

these planes are osculating with R and R. moving along the Normalization component such that arc length parameter r=r(s) with s=$|| \int_a^t \left\| \frac{dr}{du} \right\| du \ or \ \frac{ds}{dt} =$

$||r||$ *with a dot over r. r. is differentiation with respect to time r' differentiation with respect to*

distance.r.=r'=dr/dt,r..=r"=d2r/dt,r...=r'''=d3r/dt. Mapping t to s has inverse relation

of s to t given by t=psi prime(s) dt/ds =psi prime(s)=1/||r dot||The moving frame is

such that T(unit principle normal)r'=(dx/ds,dy/ds,dz/ds)The Binormal vector with

a curve B=TxN for the cross product shows all plane curves have a principal normal

and these curves are not paralle.N=(-

$sin\theta, cos\theta, 0) plane \ z = 0 \ if \ T =$

$(cos\theta, sin\theta, 0) The \ Moving \ frame \ at \ any \ point \ r. isn't zero \ but \ may \ be \ episoln \ and \ r.r: aren't zer\not p$

zero then T=r(dot)/||r(dot)||and

N=$\varepsilon(r.r.)(r..) - (r.r..)r./||r'||||r.xr..|$ |*where x is the cross product*

This moving frame in the form of a triad moves continuous along C where C is
analogous to r is the formula.:
s=r
θ *of a sphere divided into an infinite number of moving cuts or planes and each C and N are n*

Are mutually orthogonal the triplet of unit vectors T ,N and B constitute a right

handed system of basis elements E3 and cover space-time curvature as -1/2e-in cot

theta .The triad of T,N and B moves continuously along C and is the moving frame or

triad whereby T and N are the osculating plane.(touching unit tangent vectors)12

Now considering compactification of a near infinite number of dimensions or cuts in

a circle or sphere ,the area can form a point with infinite curvature moving at the

rate of the Hubble Expansion factor with the force or energy of the Big Bang.

Therefore the infinite intersections of planes in motion outward with a curvature

metric going from infinity as a point to space-time curvature metric from any mass

as derived from Einstein these dimensions which are the intersection of two moving

planes along the vectors T ,N and B will compactify below Planck Length or 10-33

cm which is the size of a string leaving at least 10 or possibly 11 non-compactified

dimensions as in supergravity or string theory rather than the 26 non-compactified

dimensions .The creation and annihilation operators which are holo- morphic and

anti-holomorphic can be used to show space-time crunching to an infinite curvature

point in a"Big Crunch" from an expanding and rotating sphere and expanded and rotating from a point of infinite curvature to this universe with all infinite

dimensions compressed in the infinite curvature point to where space approaches

nothing without reaching it and the infinite dimensions all MUTUALLY INTERSECT

AT THAT POINT .Again the compactified form of II a string theory becomes a two

dimensional sphere or circle which corresponds to that infinite curvature point or

expands with two dimensional components such as closed strings to THE WORLD

SHEET.

A complex idea being made simple is based on Ockham's (or Occum 's) Razor. All

things being equal ,the simplest explanation tens to be the right one. If a two

dimensional sphere is a circle which is the COMPACTIFIED FORM OF II a String

theory then the equation

s=r

θ *where s and $\theta$are subdivided down to infinitely small slices but with a radius that approache*

Infinite length has a compactified infinite number of dimensions of N-branes
approaches the D-infinity Brane where all of the dimensions are compactified in the
quantum bubble the dimensionality approaches the D-0-Brane just as

2

πradians approaches 0 radians on a circle but never reaches it. The radius if approaching infi

Infinite length going across 360 degrees or 2(pi) radians could radiates flat matter (2 dimensional matter or radiation)at v approaches the speed of light boundary where space-time shrinks because the inertial mass of matter approaches infinity .A radiating energy source over 2 pi radians from the quantum bubble as the radius before the "Big Bang" would travel at v approaches c. Length=Length (0)(1-v2/c2)1/2 from the Lorenzian Transformations for matter .There is length contraction at v=c just as there is time dilation where time slows down towards infinity. There is a theory that length was the first dimension de-compactified and would have to be infinite ®R from which the Hubble expansion coefficient would follow but according to the Lorenz Transformations radiation would have to be infinitely short if traveling at v=3x10^8 meters/sec which can happen in collapsed space-time where time is dilated towards infinity and space is compactified to the size of a quantum bubble or 10-33cm(String Sized).Based on this the Grand Unification Energy GUT would explode at v=c while space-time exceeds it to accommodate the mass and energy. Many questions can be answered by the arc length equation

s=r

θ and this is a very simple relationship which can be utilized for the compactific

ation of type II a string theory to aid in the explanation of the BIGBANG.

If arc length or S and theta represent space-time and R or the radius can incorporate everything else it is possible to show space-time=space/massxc2 relates to s=r(theta).The N-branes where N approaches infinity=D-0-Branes+ or –Neven Schwarz 5-brane-1/2NS-NS 2 potentials associated with Fermionic Field intensity caused by the infinite momentum limit 1/R where the Fermionic Field is gravity or the curvature of space-time caused by R ab which is the Ricci Tensor associated with the metric g ab where the infinite momemtum limit is 1/R or infinity.As the D-0-branes=0 and the gravity space-time curvature metric is infinity when dealing with the infinite curvature of a point where the ferminonic field and bosonic field have 128 permutations each acting on 256 permutations of Riemannian or Mintkowski space-time you get $0/1/\infty$ at the infinite momentum limit where R approaches infinity you get 0/0 as 1Based on this is

ther$1/\infty$ is zero therefore $\frac{0}{0}$ is mathematically everything including zero.

If the radius of s=rθ is $\frac{0}{0}$ or everything including zero and s and θ are space –
time there is an argument that this equation could be used for the infinite dimension equatio

and to mathematically explain "The Big Bang" another way if the compactified form of IIa string theory is a circle.

CHAPTER NINE
HOW   WILL THIS UNIVERSE END?

There is a theory that this universe will end with "Heat Death" whereby the accelerated expansion with progressively decreasing rotation of space-time with the approximately 750 billion galaxies will slow down and eventually stop with the distances such that any gravitational effect caused by galaxies will become negligible and space-time will approach asymptotic flatness except for the effects of the Cosmologic Constant ^ suggestive of reciprocal curvature of space-time caused by Dark Energy almost cancels space-time curvature caused by the sum total of all the mass of each galaxy. When this happens eventually stars will cool down and collapse into black holes where the majority of gravity will be sequestered and the ambient temperature will approach 2.74 degrees kelvin resulting in widespread freezing.

 An alternative to this is the "popped balloon scenario"of space-time which may possibly trigger a massive implosion and a"Big Crunch". The speed 3x10^8 meters/sec or approximately light speed"c" is a boundary in which ordinary matter can't seem to breach. Tachyons as subatomic particles seem to be able to breach this boundary as well as going backwards in time(time's arrow reversed). This occurrence can occur if tachyons are bosons with a negative instead of positive mass which would make it impossible to tachyons to travel below the speed of light at which point they would be bosons traveling forward in time and with a positive mass. It has been postulated by this author and Steven Hawking that time's arrow can reverse in a massive implosion or "Big Crunch" or by this author that negative mass as may occur at the event horizon of black holes causing a superimposition of regions in Schwarzchild Space-time netting a solution to the Hawking Paradox as mentioned in a previous chapter. Normally in a Big Crunch which could have a slow phase followed by a pop inward of approximately Planck Time 10-43 there may just initially be a light Doppler shift toward the ultraviolet instead of the infrared showing the accelerated expansion of galaxies away from each other slowing .Note this can also happen in the early state of "Heat Death" therefore measurements would have to show a gravitational metric or curvature of space-time more than canceling the anti-gravitational effect of Dark Energy and the cosmologic constant to warrant consideration about a "Big Crunch" unless the "popped balloon scenario occurs in which case the implosion could be so fast that it would equal but be opposite the explosion of the "Big Bang" with gravity instead of anti-gravity predominating.

    Space-time curves inwardly during the singularity of a "Big Crunch" as it would at the event horizon of a black hole .If a "Big Crunch" would occur the end of this space-time manifold could occur by a rip or tear in space-time as the expansion at a

progressively increasing rate and rotation at a progressively decreasing rate since "The Big Bang" .As previously mentioned the quantum bubble with infinite curvature space-time progressively uncoils towards asymptotic flatness .According to Kurt Godel's rotating universe which is shown by the clumpiness of the WOMP showing the BMR or baseline microwave radiation from "The Big Bang" the clumpiness shows increased uptake and diameter which indicates strong perturbations from the rotational vector of the expansion of the universe. Note that when you take an osculating plane that's rotating and expanding simultaneously and you take a cut of this plane the result is a spiral configuration which is what Albert Einstein postulated as the configuration of space-time in 1913.Note that in an increasing expansion with decreasing rotation space-time can achieve a rip or tear like an inflating pair of pants with a tear that becomes increasingly larger until it pops like a balloon which is a logical conclusion to Cosmic Inflation which is Dr .Alan Guth's 12 hypothesis for the expansion of space-time with conformal gravity. This tear occurs at "c" which is the boundary of the speed of light where dilated time acts as if it stops. The Lorenzian Transformation shows time dilated towards infinity or becomes infinitely long as "c" is approached closer and closer. This lack of space-time at the speed of light barrier reflects the small rip or tear in space-time with inertia being caused by the near infinite mass of the matter approaching the speed of light boundary and the resistance of space-time above light speed causes the rip which will in time increase in dimension until a "Big Crunch" s induced .During a "Big Crunch "slow phase time will slow down then stop and reverse. This only occurs if space-time travels at faster than light and created this boundary such that matter is pushing against a brick wall formed by space-time forming the near infinite inertia shown by the Lorenzian Transformation.It was also postulated that "the speed of gravity" is greater than "c" or the speed of light ,but as gravity is the curvature of space-mass caused by mass, gravity reflects the rate of curvature of space-time caused by the mass of the matter approaching the speed of light .As inertial mass for any matter approaching light speed approaches infinity the curvature of space-time at the speed of light boundary approaches infinite curvature or a point such as the tip of a cone at v=c where the cause is the pressure of space-time above v=c. If space-time exists at v is greater than c but shrinks to an infinite curvature point at v=c and then exists with the space-time curvature metric of Einstein at v <c; the speed of light barrier is a weakness of space-time that is similar to a small tear which can increase by the expansion of space-time.
It is much less likely that a "Big Crunch" will occur rather than "Heat Death" because the amount of mass versus space in this universe is very small except at the speed of light boundary or at the event horizons of black holes. If the number of black holes were to increase geometrically due to the implosion of many galaxies without those black holes "evaporating" the the amount of mass would increase compared to pinched off space-time at every event horizon. As it is there many be tears in the inflating balloon of space-time at each black hole event horizon. Despite this the space(3 space) vs mass(matter and antimatter) is so skewed toward empty 3 space (as compared with 4 space or space-time)that the curvature of space-time caused by the extant mass of all 750 billion galaxies +sum of all black holes wouldn't cause sufficient curvature of space-time to cause significant rips or tears ;and since the

ratio of 3 space to mass would increase as long as the acceleration of galaxies continue over time or in space-time the likelihood of a "Big Crunch" would decrease vs. Heat Death. If space were rife with black holes (active or dry)and if space-time was continuously collapsing in multiple spirals then the gravitational curvature metric would exceed that of Dark Energy and increase the chance of ripping of space-time unless all matter approaches 3x10^8meters/sec which is "c".

Of course with
8
$\frac{\pi G}{c4}$ *as the gravitational coupling constant approaching the cosmologic constant* ^

The chances of a Big Crunch would approach 50:50.That's 8(pi)G/c^4.It is highly unlikely that a singularity caused by non-natural causes will cause a "Big Crunch" with our level of knowledge and technology .As the Omega point(everything that is learnable had been learned)is reached ,the ability to either accidentally or deliberately cause a singularity would increase however with the Omega Point wisdom(knowledge plus experience)should prevent such an occurrence, although according to Quantum Mechanics there is a quantum state where a singularity will occur either accidentally or deliberately. This is delving on the realm of philosophy however.

As mentioned previously ,during "a Big Crunch" time will slow down ,stop or and reverse time's arrow as one goes from a greater to less entropic state reversing the Second Law of Thermodynamics during this singularity. Time's Arrow will reverse but it will still seems like it's going forward to everything and everybody within the system that the time is reversing in. It would be like saying that his body clock is synchronized for time's arrow to be pointing backwards not forward as Steven Hawking postulated happens in a Big Crunch and what this author agrees with .In this case time will point backwards until it reverts to time=0 at the point of the Big Crunch when space-time expresses infinite curvature as it does before The Big Bang and another Big Bang will follow with a heavily rotated component uncoiling in a swirl with a ballooning expansion as with cosmic inflation and the process will repeat. Again with "Heat Death"in the expansion also progressively decreases as does rotation until everything stops moving and everything freezes. This scenerio would be much more likely if there were no boundaries such as 0 degrees kelvin,"c"the speed of light boundary and spacelessness which can only be breached during singularieis. Sadly we will never live to know because the early stages of a "Big Crunch" may only be be shown by a Doppler Shift which is only slightly less to the infra-red and clocks even with gravity may not slow down a measurable degree because the measuring device is part of what's being measured and heavy masses will also dilate or slow time down which would skew the readings   .These phenomena follow the equation $\mathbb{R}a\ b\ c\ d=R\ a\ b\ c-1/2\ R\ g\ ab/R\ ab\ (1/c^2)$where space-time is curved inward. Space-time is pulled outward by inflation or with H(the Hubble expansion coefficient)with slight rotation leading to a possible swiss cheese effect in our space-time fabric in which according to Guth and others 14 other universes can form with the same or different physical laws ;however the WOMP doesn't clearly show a swiss cheese effect in space's BMR. In this cast the equation of everything becomes $\mathbb{R}$ a b c d=R a b c+1/2R g ab/R a b times $1/c^2$ instead of -1/2 R g a b because space-time is curved outward rather than inward. Also it is an axiom that each and every intelligent observer must rust his or her perception of the information gathered by the measuring device and that each measurement is based on the Heisenberg Uncetainty Principle which state measurement ranges for everything measured as the measuring device affects what is measured changing it. The is also the information exchanges of relating to "Spooky Action at a Distance"and quantum states or levels changing polarity or information at vast distances. It is also an assumption that the observer exists.

CHAPTER TEN

WHAT OCCURRED BEFORE THE BIG BANG AND WHAT CAUSED IT?

The accepted theory was that time and space began at 10-43 seconds before "The Big Bang" also known as the "Big Bang ex nihilo". This theory   opposes Newton's Law "Every action must have an equal but opposite reaction as well as "The Law of Conservation of Energy" which opposes the spontaneous appearance of a quantum bubble out of nothing (including space).

It seems logical that an implosion of another universe could have occurred prior to the explosion of the "Big Bang" in which times arrow was reversed during the implosion of the previous "Big Crunch "at which time instant time almost stopped then reversed time's arrow from backwards to forwards.
There may have been a massive Higgs Energy field which converted a quantum bubble at 10-^{33}com to matter according to be variation of energy=mass c^2 and D-0-Branes may have also existed where the zero dimensional state approximated the infinite dimensional state where the infinite dimensions were compactified or curled up with a quantum foam or something else. The multi-verse idea 15 with a n implosion preceding the explosion of "The Big Bang" seems possible.

 It is also possible that three universes collided where one was predominantly matter ,one predominantly anti-matter and and one a vacuum universe containing energy only. This may be an unlikely event however quantum mechanics states that all permutations will occur including this one. In the anti-matter universe time moves forward and with matter time moves backwards .In the matter universe time moves forward   due to positive mass and antimatter time moves backwards due to negative mass. In the "vacuum universe "there is space due to the energy of possibly the Higgs Field and has mass due to the Higgs Bosons which are massive and almost equal matter and anti-matter   ,however there is slightly more matter than antimatter. As a result of the three way collision   ,the energy of the Higgs Field of the "Vacuum universe "triggers the implosion of the matter and anti-matter universes triggering a reversal of times arrow and imploding the mass of the

antimatter and matter down to a quantum bubble which is approximately string sized which will subsequently explode 10^-43 seconds later into "The Big Bang" with initially high rotational vector at maximum magnitude just at Planck's Time then slowing at a geometric progression while the expansion or inflation occur with "The Big Bang" moving the rotating 750 billion strings into approximately 6.75x10^34 erg of energy in a 360 degree of
2
πRadian orb blast with a rotational vector ω and the Hubble Expansion CoeffiicentH^.

Collisions of galaxies occur infrequently but they do occur ,as there is a theory that Andromeda is approaching The Milky Way for a collision in several billion or perhaps many many million years ,so the collision of universes can and must occur also therefore the term "Universe" is a misnomer but is rather a "Multiverse". Matter universes exist ,anti-matter universes must exist because antimatter exists. Pure energy universes must exist with the Higgs Field because the Higgs Field exists. As all of these axioms are true the collision can occur and must occur at time "t" though highly infrequent. An since this scenario will cause the quantum bubble to occur and since a vacuum universe will force an implosion rather an explosion, this hypothesis seems very likely. Also the area   between membranes of universes has to exist it can be called or referred to as "ether" ,which would include D-0-branes which merge with n-branes in a matrix of Quantum Dots. This explains "The Big Bang" or swirl and what precedes it where the Big Bang isn't out of nothing.

From this author's first book "Megaphysics,A New Look at the Universe"it was postulated that two two anisotropic manifolds of ten dimensions each had a singularity with a dimensional reconfiguration of two six-dimensional manifolds and two four dimensional manifolds which resulted in a twetnty dimensional manifold of different configuration than the two initial anisotropic manifolds or surfaces which were unstable. With regard to theories regarding the origin of this universe Michio Kaku's idea that a 10 dimensional syupersymmetric anisotropic universe had a "popped bed sheet" effect with the popped bedsheet being the six dimensional Calabi Yau Manifold which was string sized and the four dimensional superstring cosmic universe made more sense than the "Big Bang ex nihilo".15In my theory that alternates from the three way collision theory there were two ten dimensional manifolds totaling twenty dimensions and a six dimensional Calabi Yau manifoldor something else possibly related to the Higgs Boson with the Higgs Field acting as "ether "cohabitating a "false vacuum" containing infinite space and energy from the Higgs Field. In this scenario a Higgs' Field Universe ,matter universe and antimatter universe didn't have a three way collision nor were there an almost infinite number of multiverses ,although there still were an infinite number of dimensions primarily forming D-0-branes where the dimensions were compactified from infinity to zero in a quantum foam or quantum dots in the "ether" with the two Calabi Yau manifolds and two 4 space superstring manifolds forming twenty dimensions with a huge six dimensional Calabi Yau manifold forming the other six dimensions of the non-compactified 26 dimensions of string theory cohabitating the quantum dots of D-0-branes.

The twenty six dimensions indicated by string theory included Ramajian's magic number of nature of 24 as the most stable state of four Calabi Yau Manifolds on the dimension of time and another dimension involving the Higgs Field which incorporates into the other set of four six dimensional Calabi Yau manifolds with time moving backwards or forward depending on the properties of non-compactified space and their near infinite dimensional non-compactified quantum dots. An absolute vacuum developed in an infinitely short period of time(time approaches zero but doesn't reach it)and a tidal wave of near infinite space region engulfed space-time and its contents creating a huge amount of vacuum energy from a singularity according to the equation E=mc^2 which acted as a PRIMER or catalyst either for the Higgs Field to act on the three way collision of the matter, antimatter and Higgs Field Universe to form the quantum bubble or the mass formed torn or ripped the symmetry of the recombined two ten dimensional super symmteric anisotropic universes to form an antimatter quantum bubble a matter quantum bubble both 10-33 cm with one 12 dimensional forming two Calabi Yau manifolds and two 8 dimensional forming two quantum bubbles which each contain 4 space with opposite spin2 vectors ;one clockwise one counterclockwise forming the rotational vectors of the initial 10-43 seconds before the "Big Bang" or swirl. The eight dimensional string bubbles had a crushing Ricci Tensor which was opposed by the anti-particle anti-particle repulsion twisting it into a double torus configuration which is more stable than the 8 dimensional complex causing rotation to occur greater and greater at the ends of the bubble(s)where the rotations were in opposite directions(differential spin)until the double torus acted like a pretzel and finally divided in in the middle of the two four dimensional quantum bubbles rotating in opposite directions with most of the mass occurring at the ends or outer edges and finally reaching a velocity to exceed quantum gravity of the two more stable isotropic four dimensional universes leaving two 6 dimensional string universes .The bulk of the mass spun out like a potter's wheel toward infinite length but remained string sized with Calabi Yau Manifolds at 10-33 cm. Whichever of these ideas is the simplest according to Ockham's Razor (Occum's Razor) is the most likely but again with quantum mechanics there is a probability that both occurred .In any case,"The Big Bang ex nihilio "doesn't seem likely in terms of Newton's Third Law and The Law of Conservation of Energy and Momentum.

CHAPTER ELEVEN

DOES NOTHING EXIST?

By definition nothing means non-existance.Mathematically space-time/space-time is described by the equation -1/2e^in
$\cot\theta$ divided by $-$
$\frac{1}{2e^i n}\cot\theta$. In the expression $i\,n\cot theta$ where theta is π radians you get $i\infty\cot\theta$ or $-$
$1(i\infty)$ if $\theta = \pi$ radians $- \frac{1}{2e^\infty for}n$ dimensions where $n \rightarrow$
$\infty - \frac{1}{2}e$ to the infinity power is $=$

$\infty.$ *So as space* $-$ *time is* $-\frac{1}{2e}-$
$i\, n\, \cot\pi$ *is* $\infty.$ *The expression* $\frac{\infty}{\infty}$ *is everything except* $0(zero).$ *So as the* $-$
$\frac{1}{2's}$ *cancel out e* $-$
$i\, n \cot\theta$ *where* $n \rightarrow$
∞ *gives* ∞ *in the numerator and denominator plus THE IDENTITIY POSTULATE IS MET SO WI*

INFINITE SPACE-TIME IF THERE ARE INFINITE DIMENSIONS APPROACHED IN THE MULTIVERSE.As space-time is never stationary except in a spaceless absolute vacuum due to space-time's motion(eventhough it goes below Planck Length at the Event Horizon of a black hole or pre-Big Bang or post Big Crunch)the infinite sum of space-time ∞ *and must be* $\infty.$ $\Pi(eigenstates\ 0\ to\ \infty\ of\ a1 \rightarrow an\ where\ a =$ *space* $-$
time diverges to $\infty.$ $\frac{Again\infty}{\infty}$ *is everything except* $0.$ *If the expression was* $\frac{0}{0}$ *then* 0 *would have bee*

Included. But as zero spacetime is not the infinite sum of spacetime the former MUST BE THE CASE.The exception would be if all space-time moves in equal and opposite direction with the same magnitude always..As the Riemann Metric for space-time with its 256 permutations equal zero net and the Bianchi Identity cancels out space-time in equal and opposite directions in many covariant and contravariant tensors there is an argument that spacetime(net infinite sum)/spacetime(net infinite sum) is 0/0 which includes the null set or spacelessness.This is true as 0/0 includes 0 as a solution.As the meat of the question is does all space-time as an infinite sum of the metric of R a b c d equal zero.According to Riemann it does. This indicate the mirror space-time manifolds of -1/2 e ^I n cot theta and -1/2 e –in cot theta as the left and right handed components of space-time in the Big Swirl and Anti-Swirl mentioned in this author's first book "Megaphysics ,A New Look at the Universe".Ths is true in the n-dimensional state where n approaches infinity with regard to dimensions. Although space is a container by definition and a container can exist without contents this seems to go contrary to logic although Einstein apparently stated that mass creates space but this precluded "massless "energy fields with photons which while they have a positive moving mass have a zero resting mass according to sources. Spacelessness is a boundary which if breached can implode a universe and possibly the entire space-time continuum reverted to a Higgs Field of background energy and quantum dots which are infinite non-compactified dimensions in a strict lattice formation as described by the 2 dimensional lattice equation based on the critical exponent and the act of bosons on a circle which is a compactified form of type IIa string theory or M Theory.The spherical model 17 if solvable in the presence of a field such as the Higgs Field .The spin takes on real and imaginary values and interacts with all the spins of the quantum dot lattice .It is subject to the constraint $\Sigma a = 1\ to\ N\ \sigma^2$ *where* σ *a is the first eigenstate of the sum of sigma squared =* $N.$ *Couples that with the thermodynamic limit of* $G(Gibbs\ Free\ Energy) =$

$$-kT/$$
$$2\pi \int_0^\infty F(\theta)d\theta \text{ based on the Boltzman Equation of States of matter interacting with energy}$$

and theta becomes infinitely small we develop from the two dimensional Ising
Model a two dimensional statistical model where a paritition function
Z=
$$\Sigma n\, e\left[-\frac{E(n)}{kT}\right\} \text{ where } E(n) \text{ is the energy of the nth state k is the Boltzman constant and T represe:}$$

the temperature down to zero degrees kelvin. The free energy F=-kT ln Z at
criticality correlation functions between spins
σi and σj develop a metric g ij. The expression $g\,ij =\, <\sigma i\sigma j> - <\sigma i><\sigma j>$
depending on the distance x or r spearating the states. The correlation length ε becomes ∞

At a phase transition from one state to another and at large distances where x or r
approaches infinity g i j approaches x-te-x/ε(.$\Delta = conformal\ weight$)Thus $g\,ij =$
$x^d + 2 - \eta =$
$x^{-2\Delta}$ where these are approximations. While η is the critical exponent of the field the energy op.
16.Phase transitions on a quantum dot level approach the Ising Spherical Model 18
energy
operator ϵ as a product ogf two fields whereby $\epsilon n =$
$\sigma n\sigma n + 1$ then it follows that $<\epsilon n\epsilon 0>$

equals x(any observable or expectation value $^{-2\left(d-\frac{1}{v}\right)}$ has an infinite correlation length at a ι
at a critical temperature which was already mentioned. The interaction of a free
boson or Higgs Field on a quantum dot matrix may follow the Ising Spherical Model
as the circle is compactified type II a sting theory. If spaceless ness occurs a fracture
will occur in the quantum dot lattice upsetting the
spins
$\sigma 2^2$ breaking the continuum N relating to the sum of spins. This state is in the zero external

or fermionic (vacuum state).This is shown by the Grassmann Oddvariable
$\psi - n^{2|0>}$ with the nth Fermionic Oscillator trace being $\psi - n|0 > and\ vacuum\ state$.

This describes the quantum dot lattice in terms of two dimensional open or closed
strings using the Ising Model 19. As the Fermionic vacuum state and quantum dots
are not nothing and the spacelessness state actually can be subdivided down toward
infinity where an infinite number of dimensions equals zero dimensions as in the D-
0-Brane spacelessness doesn't exist.

2. ENTANGLEMENT AND EQUATION OF EVERYTHING

FINAL EQUATION ; $\Gamma\, abcd\ \phi(|n|) > \ = \ \Lambda\ + R\, abc + \frac{1}{2} R\, g\, ab + \frac{1}{2} e^{i\pi} R\, g\, ab \div i\hbar\, \rho\, ab \ldots \ldots$

The Christoffell Symbol represents the derivative of the tensor of the fourth degree

R abcd presenting Riemannian or Lorenzian curved space-time acting upon or being

acted upon by n eigen-states of energy going from the cosmologic constant to Rayo's

number or the Grand Unification energy of 10^19GEV or 10^77 joules being

entangled in a non-chaotic arrangement of Spooky Action at a Distance and the

interaction of strings according to type II and IIA string theory as well as the

Heterotic 8x8 string theory and the SO32 string theory forming M theory

formulating the Superbrane consisting of 524,288 permutations of energy levels as

eigenstate. The expression R abcd(n) where n is the number of eigenstates of energy

going from the Cosmologic Constant to the Grand Unification Energy acting upon or

being acted upon by curved space-time being curved by the mass equivalent of the

separate disparate energy states is represented by the Christoffel symbol as the

derivative of the tensor of the fourth degree of space-time .The term psi(|n|)>

represents the entanglement of these myriad disparate energy states acting upon

each level of space-time forming the derivative of the tensor of the fourth degree of

space-time. In actually Entanglement exists to some degree in every energy state

except the ground state as represented by The Cosmologic Constant so when the

derivative of the tensor of space-time is acted upon by the cosmologic constant the

constant vanishes or becomes zero as the derivative of any constant is zero. Of

course

$e^{\wedge}\pi i$ *is* −

1 *according to Euler's Identitiy making the tensor expression for antigravity* +

$\frac{1}{2R}$ *g ab(flattening space* − *time)and gravity curving space* −

time inward as the Black Holes.

Of course the ascending and descending spheres or geodesics of space-time going

from a point of infinite curvature to flat space-time of infinite diameter are

represented by $2\pi R$ where 2^n+1

π is in the numerator and $2^{n\pi}$ *is in the denominator making the spiral operator acting upon an*

Angular momementum for the geodesics increasing flatness and decreasing to a

point of infinite curvature.The relates to the Stress Energy Tensor time the

gravitational coupling constant or

8

$\frac{\pi G}{c4}$ $T\,ab$ in the numerator and $\frac{8\pi G}{c4}T\,ba$ in the denominaator. As the ground state is the cosmolc

Cosmologic constant
or
Λ from the First Event Time Oscillation Paradox Theory the stress energy tensors cancel leav

Leaving R as the cosmologic constant and the expression
$2\pi\,\Lambda$ where R is the cosmologic constant and $2^n +$
$\frac{1}{2^n leaves}$ 2 pi times the ground state energy level whereby 2 pi is the circumference of the first

Geodesic of space-time where the radius is the ground state energy level or the
cosmologic constant.

$\Gamma\,abcd\,\phi(n) >$
$= 2\pi\,\Lambda\, +\; R\,abc\, +\frac{1}{2}R\,g\,ab +\frac{1}{2}\; e^{i\pi}\; R\,g\,ab \div\; i\hbar\;\rho\;ab\;$ where $\hbar$

$= h\frac{}{2\pi}\;$ making the expression compactify to a circle with space
$-$ time as the circumference and the Riemann Forces of Nature as the positive and negative r

Radiating grom the center of the compactified circle. This is the expression space-
time=space/mass.

BIBLIOGRAPHY

1:Peat,F.David. Superstrings and the Search for the Theory of Everything.Yang Mills Forces p.114

2.Kaku,Michio. Strings,Conformal Fields and M theory.Ising Model p.176-78
3 : Wald ,Robert .General Relativity.Chicago,Ill. University of Chicago Press.1984 4.CPT THEOREM; Quantum Field Theory Kaku ,Michio
4:Metric tensor(General Relativity)Wikipediaa and Spacetime.en.m.wikipedia.org.spiral Space-time Einstein 1912 Fractal Time.p.108-109 Braden ,Gregg 2009 Library of Congress HAWKING RADIATION. Wikipedia
5.Peat,F.David .Superstirngs and the Search for the Theory of Everything.p.106-107.Calabi Yau Manifolds

6.Kaku,Michio .Quantum Field Theory. Renormalization Actions in Quantum Field Theory
 7.Peat,F.David .Superstrings and the Search for the Theory of Everything.
8.Kay, David C. Tensor Calculusp.129 Osculating Plane

9.Kaku, Michio. Strings,Conformal Fields and M theory.

10.Peebles,P.J.E .Principles of Physical Cosmology.p

11Chang,Alan. HAMILTON JACOBI EQUATIONS UNIVERSITY OF CHICAGO 2013.Zeno's paradox: The Math Forum at Drexel University

12:Tipler,Frank j.The Physics of Immortality

13.Godel,Kurt.Godel's Incompleteness Theorems en.m.wikipedia.org

16:ibid item#7 p.237-42

14.Randall,Lisa. Warped Passages
 15:Green,BrianThe Elegant Universe.
and
14.Wick,Mitchell Albert .Megaphysics ,A New Look at the
Universe.

15:Kay,David.CTensor Calculus.
18:Kaku,Michio. Strings, Conformal Fields and M Theory.
19:Wikipedia.Electronen.wikipedia.org/wiki/Electron
20:Spooky Action at a Distance Quantum Entanglement
Wikipedia. Or en.wikipedia.org/wiki/Quantum entanglement
 21:Hau,Len. Harvard Research circa 2003.
BIBLIOGRAPHY

Barrero,John D.The Anthropic Cosmological Principle.Oxford
England.Oxford Press.1986
Brade,Gregg.fractal Time 2009 Library of Congress.
Greene,Brian.The Elegant Universe.NewYork.Vintage Books
editor Random Press.1999
Hawking,Steven and Penrose,Roger.The Nature of Space and
TimePrinceton,N.J;Princeton Science Library 1996
Kaku,Michio.Quantum Field Theory.A Modern Introduction.Oxford
university Press.1993
Kaku,Michio.Strings,Conformal Fiels,and M theory 2nd
edition.Springer Press.2000.
Kay,David C.Tensor Calulus Schaum's Outline Series.
N.Y.McGraw Hill 1998.
Peat,F.David.Superstrings and the Search for the Theory of
Everything.Chicago.Contemporary Books 1998
Peebles,P.J.E.Principles of Physical Cosmology.Princeton Series
in Physics.Princeton University Press 1993
Wald,Robert m.General Relativity.Chicago,Illinose.University of
Chicago Press 1984
Wikipedia: on lin encyclopedia.
Randall, .Lisa .Warped Passages HarperCollins
Publishers.N.Y.2005
Tipler ,Frank J. Physics and Immortality. Anchor Books division
of Random House.1993

16:ibid item#7 p.237-42

16:ibid item#7 p.237-42

GLOSSARY

Abelian :equations having a coefficient or variety in a specific group,g,,algebraic number fields,tensors of the same degree or cohominy group

Anisotropic :not isotropic,lacking observational symmetyry

Anti-symmetric:tensors or vectors that are equal but opposite and can therefore partially cancel or cancel

Aymptotic:that which approaches a level or degree but never reaches it;asymptotic flatness appears without curvature but doesn't reach it

Bianchi's Identity: The identity of groups of Riemannian 4 space that is anti-symmetric and Abelian and cancels each other out of being equal but opposite
"The Big Bang"A theory proposed describing a Friendman type I open expanding f;at universe with is homogeneous and isotropic

"The Big Swirl "A Big Bang with a progressively decreasing rotational vector from an infinite curvature point of space-time to asymptotic flattness

Black Hole :collapsed matter from a neutron star or galaxy with extreme curvature of space-time at the central nexus due to extreme gravity of of the spiral space-time

Calabi Yau Manifold:a surface which represents a relative isotropic portion of space-time with a puckering to accommodate multiple dimensions considered a twisted variant of the orbifold
Choas:absolute disorder

Chiral:a mirror image or absolute symmetry

Closed string:a two or one dimensional building block of matter from energywith movements in 10 or 26 dimensions without breaking the string

Compactified:when every point of the dimensions are curled up mathematically making the size approach zero.First determined by Kaluza and Klein

Conformal Space:when every point in space relative to every other point maintains its relative position regardless of what the space is doing

Dark Matter;an indirectly measured mass causing perturbations in gravity(the curvature of space-time)caused by mass.Acts as cosmic glue containing possibly baryonic particles and neutrinos

Glossary page 2

Event Horizon:area where a black hole is perceived by measurementsEntropy:degree of disorder

Entropy:degree of disorder
Ex nihilo:out of nothing

M(Membrane)theory:the 5 dual string theories into one massive theory of everything which incorporates membranes which vibrate and incorporate all energy and matter

Isoropic:observational symmetry

Geodesic:a unit of space-time
Gravity:the curvature of space-time caused by mass;actually an effect not a force

Membranes:a description of matter in terms of energy states with stress energy densities described in the number of states with regard to dimensions

N;number of dimensions in N dimensional space

Open string:a two or one dimensional bulding block of matter with movements in a multidimensional plane

Orbifold:space-time manifold in an open twisted cone configuration utilized in string theory

Relativity:the behavior of matter and energy with regard to other matter and energy;energy and space have a different vantage point from other matter and energy including stress energy,time and mass with changes regarding relative velocity

RicciTensor:that tensor which represents inertial mass or resistance against pull or push
Riemann Forces:all strong and weak forces in nature

Riemannian Space:Mintowski space with Riemann curvature of space-time caused by mass.Flat space if no mass is present

Scalar:the magnitude compone t of a vector or tensor with regard to direction

Six dimensional string manifold:curled up closed strings in configuration according to Kaluza and Klein which is 10-33 cm and may be Calabi Yau Manifolds

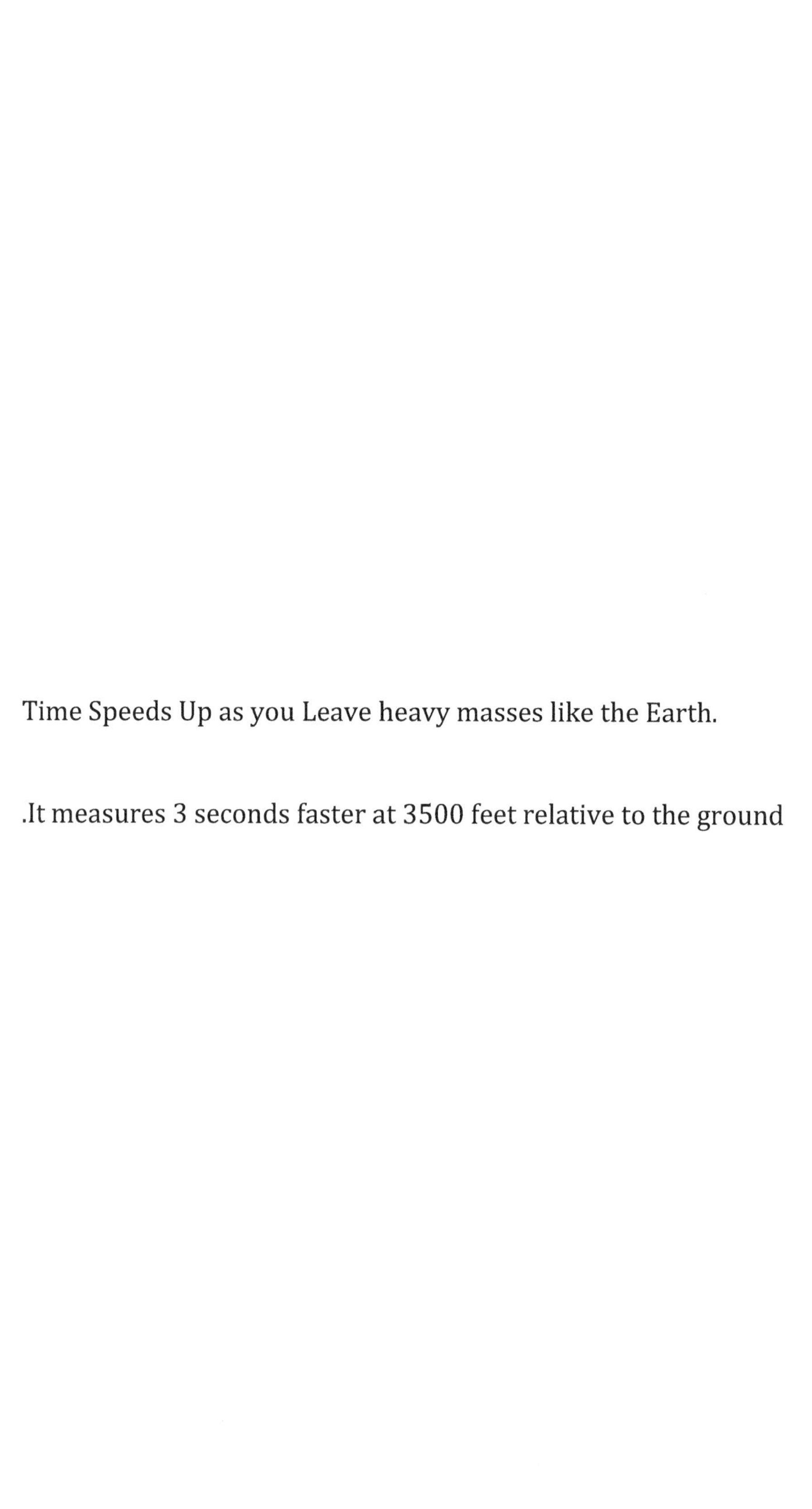

Time Speeds Up as you Leave heavy masses like the Earth.

.It measures 3 seconds faster at 3500 feet relative to the ground

THE DEATH OF A UNIVERSE IN THE SPACETIME CONTINUUM

Just as Dark Energy will break up all matter and anti-matter down to fermions

and anti-fermions it will eventually break the gluons which hold the fabric of space

together in extremely fast or constricted time aeons into the future causing Big

Crunch. All this will also dissolve dark matter or WIMPS which not only hold matter

together but if they incorporate information which travels through neutrinos it will

acts in a manner similar to decomposition in death as the matrix of dark matter

mimics that of a peripheral nervous system if it is indeed information. While this can

be determined by Q bits and each permutation involved in entanglement is a Q bit it

doesn't necessarily mean this universe is a simulation as there is still no off switch.

To have an off switch there must be a region of nothing and a hard barrier when in

fact there are gradients and nothing doesn't exist as quantum fields always exist as

do leptons and gluons(space).

SPACETIME AND MASS

Precisioned clocks (atomic clocks) over different parts of the planet earth show

minor fluctuations in real-time as predicted by the relationship of space-time and

mass as described by Albert Einstein . The dimension of time dilates or slows down

as any heavy mass is approached and constricts or speeds up as one travels away

from a heavy mass to a lighter mass. The more time is dilated the more it constricts

space and the more it's constricted the more it flattens or dilates space. This goes

according to the formulae

$$8\frac{\pi G}{c4}\ T\ ab =$$

$R\ ab - \frac{1}{2}R\ g\ ab\ for\ gravity\ which\ curves\ spacetime\ inward\ and\ R\ ab +$

$\frac{1}{2}R\ g\ ab\ for\ anti-gravity\ which\ curves\ space-time\ outwardly\ or\ flattens\ it.$

These equations fit neatly into the Equation of Everything $R\ abcd = \Lambda + or -$
$R\ abc + or - \frac{1}{2}R\ g\ ab \quad \div \quad R\ ab$ where R ab=$i\hbar\rho\ ab/c$^2

THE EQUATION OF EVERYTHING

AUTHOR: DR. MITCHELL ALBERT WICK

SPACETIME=SPACE/ MASS x c

This is an addendum or sequel to book "*Megaphysics, A New Look at the Universe*" written by the same author with some new twists or insights based on the same fundamental ideas and principles.

"A TRUE SCIENTIST ASKS QUESTIONS OF HIMSELF AND ANSWERS THEM FOR THE WORLD ": - Dr. M. Wick

Table of Contents

FORWARD AND PROLOGUE

Description of background

After the publication of this author's previous book "Megaphysics, A New Look at the Universe" this author continued on the quest to continue to answer the eternal questions. Using the basic premise of the first book, that space-time curvature is described by the spiral fractal formula this author was able to postulate that the quantum mechanical formula for space-time was utilized to formulate the GUT energy of 10 19 Gigaelectronvolts with Planck's Mass for a quanta. This same basic formula was utilizable with regard to Relativity using Einstein's Law of Relativistic Gravity and tensor calculus. Both equations described the same equation Space-time=Space/massx k where k=1/c 2. This basic inch long mathematical relationship was first described in this author's book "Megaphysics, A New Look at the Universe" however a deeper exploration into the mathematical relationship unifies Quantum Mechanics and Relativity as well as describing virtually all phenomena of the multiverse.

Dedication and inspiration

This author would like to thank and acknowledge the continuing support of family and friends in this endeavor including Dr. Bert Warren (uncle), Ann Lori Tucci (sister) and James Andrew Tucci (brother in law) without whom a long arduous task only prompted by a love of nature,a desire to seek the truth and a love of mathematics was made infinitely easier.

CHAPTER 1

THE FORCES AND HOW THEY INTERACT

The strong force, electromagnetism, the weak force ,gravity and weak perturbations from quantum fluctuations are interact on space-time. As energy= mass xc 2 these forces can be reincorporated into their corresponding mass with the strong force which binds the nuclei of atoms together by quarks relating to the fission of a atomic nucleus, the weak force relates to radioactive decay over time from a heavy nucleus (such as Uranium 238 with unstable neutrons decaying into alpha and beta particles with corresponding gamma and x ray production), gravity which is the curvature of space-time due to mass.

As a result gravity is considered a weak force in Yang Mills Terminology [1] compared to electromagnetism and the Strong Force associated with fission and fusion (which occurs in stars). Electromagnetism is the transfer of electrons from covalent bonds and electrovalent solutions where the transfer of electrons relates to magnetic fields if the potential difference is great enough and the electron spins are aligned into a force sufficient enough to induce a magnetic field as described by Coloumb's Law. The field is described as the quantity of charge of each interacting pole/radius squared which is the distance between Q1 and Q2 where Q1 and Q2 are the charges involved and Coulomb's Constant is multiplied by Q1Q2. When converted to tensors the Ising Equation [2] is derived. Electromagnetism causes strong perturbations. They involve dipole moments of opposite spins in two states +1 and -1 incorporated in a lattice. In local fields H the free energy=

$$\int d \text{ to the } d \text{ power } x\{AH \text{ squared} + \Sigma i = 1 \text{ to } d \, Z \, i(\partial iH)2 + \lambda H4$$

...which represents the free energy associated with electromagnetism. The Maxwell Equat

Equations also describe electromagnetism while the Ising Equation utilizes statistical mechanics to describe the free energy with electromagnetism .The Maxwell Equations $\underline{3}$ involve Gauss's Law

$$\nabla . E =$$
$4\pi\rho$ which is related to Poisson's equation which is the derivative operator of a dual vector fie

Is 4 pi times the energy density of matter. E stands for an electric field and applied to the Poisson's equation. The Maxwell Faraday Equation is

$$\nabla x E = -\frac{\frac{1}{c}\partial B}{\partial t}\, where\ B\ relates\ to\ the\ electromagnetic\ field\ such\ that\ \nabla . B = 0.$$

The ising Model for electromagnetic fields with its associated free energy is incorporated into the YangMills GUT which is described as are all strong perturbatory forces on the left side of the quantum mechanics equation for space-time=space/mass(c2). Gravity is described by Einstein's Equation for Relativistic Gravity

$$8\pi T = R\,a\,b - \frac{1}{2}R\ g\,ab.$$

CHAPTER 2

THE ACTION OF METRICS ON SPACE-TIME

SPACE-TIME CURVATURE IS ACTED UPON BY A METRIC g relating to any mass, which is causing the curvature. The metric g curves space-time according to the mass and is described by the inertial moment R a b which is the Ricci Tensor. As a result gravity can be described as g a b or C a b that is Weyl's Conformal Tensor describing conformal gravity. According to Einstein these tensors relate to the stress energy tensor T a b as Einstein's Equation of Relativistic Gravity

$$8\,pi\ T\,a\,b = R\,a\,b\text{-}1/2R\,g\,a\,b$$

where R g a b is the gravitational metric acting on space-time causing Conformal Gravity. The equation was derived from the Lorenzian Transformations but basically it says stress energy=inertia-1/2 gravity which is zero when the stress energy tensor is zero which is pre-Big Bang or post Big Crunch.

In black holes the gravitational effect from a collapsed neutron star is so great even light can't escape. Beyond the event horizon of a black hole space-time is collapsed in a spiral or vortex configuration to almost zero and mass is collapsed to a nearly infinite density. This mimics pre Big Bang where the Einstein Tensor G a b approaches 0. This is illustrated in the Relativistic form of

$$spacetime=space/mass\ (c2)\ where\ R\,a\,b\,c\,d = R\,a\,b\,c\text{-}1/2R\,g\,a\,b/R\,a\,b$$

where space-time is curved inward by the collapsed mass in the black hole. In essence, the space-time curvature metric is gravity acting on R, which is space-time that is curved by the metric.

In the pre-Big Bang epoch (up to 10-43 seconds or Planck's Time) the space-time curvature metric approaches infinite curvature as a point has infinite curvature and the extreme mass of the pre-BigBang quantum bubble curves extremely small space-time to infinity. In this case gravity approaches half of a very large non infinite number and anti-gravity from anti-matter approaches a very large non-infinite number but inertia equals the sum total of the gravitational and anti-gravitational metric acting on space-time with almost infinite curvature. This all indicates that GRAVITY ISN'T A FORCE BUT THE EFFECT OF SPACE-TIME CURVATURE CAUSED BY ANY MASS.

THE ACTION OF ANY METRIC g IS DESCRIBED BY THE ACTION FORMULA:

$$S=-1/2k2(-g)1/2R$$

and this is the probable action of any metric in space time R with curvature metric (-g)1/2. R can be described in terms of dimensions such as d to the n dimensional power and this is extremely useful in string theory with its postulated 26 dimensions compactified to 10.

CHAPTER 3

RELATIVISTIC GRAVITY AND HOW IT INTERRACTS WITH SPACE-TIME

As previously mentioned Relativistic Gravity is described by 8(pi) T=R a b-1/2R g a b where R g a b describes Relativistic Gravity and it's effect on space-time. The space time curvature metric g a b is described in the tensor equation

$$R\ a\ b\ c\ d = R\ a\ b\ c\ -1/2\ R\ g\ a\ b/R\ a\ b$$

where R a b c d described curved Lorenzian 4 Space-time R a b c describes flat Mintkowski 5(footnote not dimensions) Space and R g a b described the space time curvature metric acting on flat Minkowski Space causing the curvature caused by the mass described by the Ricci Tensor R a b. The entire expression is multiplied by 1/c2 to give the Relativistic form of the Equation of the Universe. Utilizing Einstein's Equation of Relativistic Gravity and this author's equation of everything and solving for the stress energy tensor T a b one gets the gravitational constant G=6.67x10-11 Newton. Setting R a b from both equations equal to each other by the transtivity postulate a=b a=c therefore a=c T I j=T a b .T ba where i= initial event and j=final event and mass=energy/c2 so R a b=T I j/c2=T a b.T b a/(c2)2 as T ab.T ba=T I j/||c2||2 and 8 pi T=R a b-1.2 R g ab via Einstein's formula where 8(pi)T ab.T ba/c4 reflects stress energy tensor T I j=R ab-1/2 R g ab at 0 stress energy or the Einstein Tensor G a b ;T ab.T ba=G the gravitational constant as 8(pi)G/c4=8(pi)T ab.Tba/c4 and 8(pi)G/c4 is the gravitational coupling constant k the stress energy of 0 at constriants of pre-BigBang post-BigCrunch and at event horizons of black holes Stress energy=Gravitational constant or 6.67x10-11 newton/meter/sec2 approaches 0.The difference is attributed to weak perturbations as described by the action formula S or the Hamiltonian Operator for n eigenstates of energy as the 0 or null eigenstate is approached.

An operator is a complex function, which operates on another function such as the LaPlacean Operator, which acts as a second-degree differential equation of the function it is operating on to the upward limit of the operator. An eigen-state is a state of matter or energy with regard to the Operator that operates on another function such as the quantum level of matter with regard to energy.

CHAPTER 4

THE EQUATION OF THE UNIVERSE IN TERMS OF RELATIVITY

Albert Einstein postulated that all motion is relative and not absolute. When a mass m travels at approaching the speed of light boundary inertia approaches infinity space-time approaches 0 and length shortens to infinitely short as per the Lorenzian Transformations. As item A of mass m travels west in an environment that is traveling velocity B when item A is traveling at velocity A the total velocity is the sum of A and B unless there's another manifold (surface) which is traveling at velocity C which if positive is added to velocity A and B and negative is C is traveling less than velocity A and B. The speed of light boundary can't have anything with near infinite mass and inertia hit the boundary where spacetime =0 and since velocity is diminished by inertia even light cannot travel at this boundary but travels a near infinite speed below the boundary C.

THE EQUATION OF THE UNIVERSE is

spacetime=space/mass (c-2) or R a b c d =R a b c-1/2 R g a b /R a b with the entire expression multiplied by k which is 1/c2.

In terms of metric tensors Lorenzian curved space-time or Riemannian 4 space is described as R a b c d. R a b c described flat space on which the metric of gravity curves space into curved space-time. The metric of gravity emanating from mass m whose inertia is R a b which is the Ricci Tensor curves flat Mintowski 3 space either inward or outward depending on the mass being acted upon by the mass that's doing the acting. As a result the net space-time curvature is the tensor sum of the space being curved by the mass doing the curving and the mass being acted upon by the metric of the mass doing the curving. This sum is described by R g a b where g a b is the metric of the mass described by the RicciTensor R a b acting on R which is the space-time curvature caused by the metric g a b of mass m.Spacetime=Space/mass so Riemannian 4 space or curved Lorenzian Spacetime=Mintowski 3 space or flat Riemannian 3 space R a b c minus the tensor sum of the metric g ab acting on space R with the space time curvature metric. This is what Einstein described as gravity. The mass part of the expression spacetime=space/mass is described by the RicciTensor which describes inertia which is the resistance against push or pull by the mass described. So the expressions spacetime=spacetime and spacetime=space/mass which converts to spacetime=space-1/2 spacetime curvature metric described as gravity divided by mass or inertia as described by the Ricci Tensor.

Riemannian 4 space as a tensor of the fourth degree is a 4x4 determinant with 12 components of which the Bianchi Identity 6 cancels out 4 of the permutations of R a b c d. there can b 4 to the 4th power of permutations in Riemannian space-time, which is 256 permutations. This is includes covariant and contravariant tensors which when obeying Bianchi's Identity are Abelian and anti-symmetrical being equal and opposite in both magnitude and direction. It is beyond the scope of this text to show all the permutations of Riemannian 4 space however the mathematical relationship of Riemannian or Lorenziann 4 space and Mintowski or Riemannian 3 space is what's important to this author.

The space-time curvature metric of Einstein emanates from Einstein's Equation of Relativistic Gravity where a progressively increasing mass as described by the RicciTensor of inertia -1/2 gravity or the space-time curvature metric which also progressively increases as inertia increases progressively curves space time toward infinite curvature which is the space time of the pre-BigBang quantum bubble which may or may not be smaller than Planck's Length 10-33 cm depending on whether deSitter Space 6 hits a hard downward boundary at 10-33 cm. String Theory describes space in terms of the Orbifold which is a complex twisted cone which eventually can twist into a Calabi Yau Manifold 7or surface which is a continuous surface in a puckered appearance of a double torus that communicates with other Calabi Yau Manifolds all six dimensional and in motion .

As increasing inertia and increasing gravity do not increase at the same rate as the speed of light boundary is approached space-time progressively curves into a point at v=c. According to the Lorenzian Transformations infinite mass in dilating time and reducing space-time with progressively increasing curvature causes inertia to increase at a greater rate than gravity because the metric g a b is acting on progressively increasing space-time curvature which is R g a b and the space-time curvature metric is half the inertia because the covariant and contra-variant tensors of space-time curvature are Abelian and anti symmetric as inertia approaches infinity at v=c which is when the Einstein tensor G a b=0 which is the stress energy tensor T a b at infinite space-time curvature. Anti-symmetric tensors cancel out in magnitude but with opposite direction.

CHAPTER 5

THE EQUATION OF THE UNIVERSE IN TERMS OF QUANTUM MECHANICS

In terms of quantum mechanics spacetime=spacetime=space/massx1/c2 is described in terms of Planck's Mass which is the minimum mass a quantum particle or two quanta can occupy and is approximately 1.22x10-24 kg. Mathematically Planck's Mass is the square root of hc/8(pi)G where h=Planck's Constant 6.63x10-34 G is the gravitational constant 6.67X10-11 newton-meters/sec2 and c is the speed of light boundary. Space-time is described as the n dimensional state of a point particle x at time t as a probability function and is operated on by the Hamiltonian Operator defined as –h2/2m(the LaPlacean Operator with respect to the second derivative so utilizing the above space-time is Ha|(r,t)|2d n r where d n r is the n dimensional state of point particle r with respect to t or time|(r,t)|2 is the probability density of point r with respect to t in n dimensional space. This equals -1/2 e + or –i to the n cot theta power.here i=the square root of -1.=or – I .e is the inverse or reciprocal of in which is the integral of du/u which relates to the spiral fractal formula defining the ground state in a relativistic universe as in 1=0 and in infinity= infinity. Theta is the angle of trajectory at the Big Bang, which is pi radians or 180 degrees, to the infinity power is infinity and e –infinity power is zero. Therefore e-I to the n cotangent theta power defines the ground state on in 1=0.e I n cot theta power is a reciprocal function and describes 1/0=infinity. Therefore Planck's Mass (c2)+H a |(r,t)|2d n r where this expression multiplied by the La Placean Operator in n dimensions (note that the inverted delta sign also reflects ta momentum operator often used in non-relativistic calculations)but in this instance the Operator defines the wave function of r with respect to time t. Note also that the Hamiltonian Operator=-h2/2mLaPlacean Operator)2+Vo and reflects momentum p=mv where p=momentum. So the momentum of quanta with Planck's Mass is incorporated over n dimensions from a to n eigenstates of energy Weak perturbations must include the momenta of quanta and a to n eigenstates or energy levels. Note the Hamiltonian Operator operates with respect to a eigenstates of energy or quantum energy levels. Therefore c2 (hc/8 pi G) 1/2+H a|(r,t)|2 d n r times inverted delta to the n power=-1/2 e +or −i(square root of -1)to the n cot theta power where n is the number of dimensions. In this case the Hamiltonian operator covers weak perturbation of quanta in n dimensions and describes the wave function of point particle r with respect to time in the n dimensional state with respect to a eigenstates.c2 (hc/8 pi G)=10 19 Gigaelectron volts which is the Grand Unification Energy which includes all forces except weak perturbations from quantum fluctuations which is described by the Hamiltonian operator in a eigenstates in the n dimensional state.e i(square root of -1) to the n cot theta power is infinity. Therefore 10 19 Gev+Hamiltonian (quantum fluctuations)=infinity. This occurs in the multiverse where the n dimensional state approaches infinity and can be proven by subdividing a sphere to one second of arc or 1/3600 degree. When this one second of arc is a plane in motion the topological surface mimics a oscullating plane8;the math is beyond the scope of this text. These oscullating planes from a topological standpoint define a dimension for each oscullating 8 plane.According to Zeno's Paradox 11 each degree subdivided ad infinitum prevent two solid objects from touching. Using this a second of arc can be subdivided down an infinite number of times each being an oscullating plane implying an infinite number of dimensions. With an infinite number of dimensions for space-time the left side of the equation goes from 10 19 approximately equals infinity=infinity on the right side to infinity on the left side equals infinity on the right side and is conclusive of the multiverse. This applies to e I(square root of -1) to the n cotan theta power. This applies to everything except the ground state, which relates space-time to e -1 n cot theta power which is zero space-time on the right. On the left Planck Mass is zero in the ground state as this is the vacuum state so c2 (hc/G) 1/2=0 and in the zero dimensional state the Hamiltonian Operator or point particle|(r,t)|2 d n r times the La Placean Operator in the zero dimensional state is also zero because the derivative of zero is zero. Therefore 0=0 in the ground state and the Equation of the Universe or Multiverse is upheld with regard to Quantum Mechanics. Note that this expression equals the Relativity Expression R a b c d=R a b c-1/2 R g ab /R ab times 1/c2 because they both equal space-time=space/mass (1/c2) due to the transtivity postulate. This unites Quantum Mechanics and Relativity.

CHAPTER 6

DEMONSTRATION THAT THE QUANTUM MECHANICS AND RELATIVITY EQUATION FOR EVERYTHING ARE THE SAME EQUATION

AS MENTIONED EARLIER BOTH

$$R\ a\ b\ c\ d = R\ a\ b\ c\text{-}1/2\ R\ g\ a\ b/R\ a\ b\ times\ 1/c2$$

$$and\ c\ 2(hc/8piG)1/2 + H\ a|(r,t)|2d\ n\ r\nabla\ n = -\frac{1}{2e} + or -$$

$i\ n\ cot\theta$ *power are equal to each other and equal spacetime* $=$ $\frac{space}{mass\left(\frac{1}{c2}\right)}$. *This indicates that in black holes spacetime curves inward so* $R\ a\ b\ c\ d =$ $R\ a\ b\ c - \frac{1}{2}r\ g\ a\frac{b}{R}a\ b\ or\ R\ a\ b\ c\ d = R\ a\ b\ c - \frac{1}{2}T =$ $R\ g\ a\frac{b}{R}a\ b\frac{1}{c2}$. *Black Hole entropy according to Steven Hawking is 0.29 which approaches 0.*

This is according to Steven Hawking who stated that black holes have 252 separate states of matter and the entropy is S=2PI(Q1Q5N) 1/2 where N is the number of states or eigenstates Q1 and Q5 are the differential charge between the ist and 5th eigenstates.9.s= entropy or degree of disorder. The one-brane in M theory (Membrane Theory) is described with the monopole or a fixed negative charge suggestive of an electron and the five-brane represents 4 space or curved Lorenzian Spacetime which spirals into the black hole event horizon. As a result the energy of a monopole acting on Lorenzian curved spacetime across 252 states of matter reveals black hole entropy.S BH=A/4L2p=c3A/4Gh where A=cross sectional area L p=Planck's Length G is the Gravitational Constant and h=Planck's Constant(h has a bar across upper stem in all references to Planck's Constant.This indicates that as cross sectional area approaches 0 entropy in a black hole approaches 0(zero)according to the Bekenstein-Hawking Equation. This mimic s entropy at pre-Planck Time in the quantum bubble where a mix of 50% matter and 50% antimatter are solidified by enormous pressure into what might actually be a lattice formation much as a diamond would occur. Indeed the central locus of a black hole post event horizon may also have such a configuration, as with tremendous pressures strange matter and liquid states of matter, which would under other ambient conditions, not be liquid or solid. Before one proposed that radiation such as electromagnetic radiation can occur in a liquid or solid state in a black hole there must be substantially more empirical evidence that radiation can occur as a perfect gas remembering that photons have a measurable mass and light is bent by gravity(light can curve space-time).Incorporating space-time as a perfect fluid and radiation as a perfect gas one would anticipate quasars as spuming out with space-time as a syringe that overflows would spill out water. This would have to occur in black holes because they 'DRY UP' and evaporate over time. The Hawking Paradox couldn't explain how matter and energy would just evaporate due to the Law of Conservation of Energy and his people worked for many months on a solution. A plausible solution is that as a perfect fluid space-time "leaks" or spills out of a black hole pushing matter out with it over a 2 pi radian or 360 degree sphere with the center of the sphere having the most pressure of the "muzzle blast" just as a water hose spumes water out of a tank, the black hole is the tank; the quasar is the center of the hose and the periphery extends out over 360 degrees until it gradually dissipates or gets pulled in by space-time from the mass of another object or the same black hole.

Space-time curves in also during the singularity of a "Big Crunch". This is a likely scenario for the end of this space-time manifold or universe. Space-time is expanding at a progressively increasing rate according to the Hubble Expansion coefficient (H). In addition space-time is rotating at a progressively slow rate since the "Big Bang" when space-time in the quantum bubble had almost infinite curvature progressively uncoiling toward asymptotic flatness. According to Kurt Gödel this universe is rotating (Gödel's Rotating Universe) which is must do proven by the clumpiness of the WOMP showing the BMR or background microwave radiation from the Big Bang. The clumpiness shows increased uptake and diameter, which indicates strong perturbations from the rotational vector of the expansion of the universe. Note that when you take an osculating plane that's rotating and expanding simultaneously if you take a cut of this plane or slice the result is spiral configuration, which is what Einstein postulated as the configuration of space-time in 1913. Note that in an increasing expansion with a decreasing rotation space-time can achieve a rip or tear like an inflating pair of pants with a tear that becomes progressively larger until it pops like a balloon, which is a logical conclusion to Cosmic Inflation, which is Dr. Alan Guth's 10 hypothesis for the expansion of space-time. This tear occurs at "c" which is the boundary of the speed of light where dilated time acts as if it "stops". Why does the Lorenzian Transformation for time shows it dilates to infinity or becomes infinitely long in duration making space-time actually approach zero as c is approached closer and closer.

This lack of space-time at the speed of light barrier reflects the tear or rip in space-time with inertia being caused by the near infinite mass of the matter approaching the light speed boundary and what has been proposed as space-time travel faster than "c" but not at "c". The infinite inertia of matter approaching light speed and the resistance of space-time above light speed causes the rip which will in time increase in dimension until a "Big Crunch" is induced. During a "Big Crunch" time will slow down, stop then reverse time's arrow as one is going from a greater entropic state to a lesser entropic state; a reversal of the Second Law of Thermodynamics during this singularity. Times arrow will reverse but it will seem to be going forward to everything within the system that the time is reversing in. it would be saying that his body clock is synchronized for time's arrow to be pointing backwards not forward as Steven Hawking postulated happens in a "Big Crunch" and what this author happens to agree with. In this case time will point backwards until it reverts to time=0 at the point of the "Big Crunch" when space-time expresses infinite curvature as it does before the "Big Bang" and another "Big Bang" will follow with a heavily rotated component uncoiling in a swirl with a ballooning expansion as with cosmic inflation and the process will repeat. There is an alternate theory called "Heat Death" where the expansion progressively decreases, as does the rotation until everything almost stops moving; temperatures drop to near 2.74 degree kelvin, which is the BMR, temp from "The Big Bang" and everything freezes. This scenario is also possible but only if there were no boundaries such as 0 degrees kelvin, "c" and spacelessness which can only be breeched during singularities. Sadly we will never live to know because while the early stages of a "Big Crunch" may only be shown by a Doppler Shift away from the infra-red toward to ultraviolet our clocks even with gravity camera may not slow down a measurable degree because the measuring device is part of what's being measured. Then a critical point would be reached when the balance of the "Big Crunch" would occur in less than Planck's Time 10-43 seconds and everything would swirl into a vortex which may only be 10-33 cm or Planck's Length much like the vortex of space-time swirling into a Black Hole. This would indicate extreme gravity as curvatures of space-time would coil to a point and would have to be countered by anti-gravity or Dark Energy as it was at "The Big Bang" which involved anti-particle anti-particle repulsion. Again these phenomena follow the equation R a b c d=R a b c -1/2 R g a b/R a b times 1/c2 where space-time is curved inward. Space-time is pulled outward with Inflation or the Hubble Expansion coupled with slight rotation leading according to Alan Guth a Swiss cheese effect in our space -time fabric in which other universes can form with the same or different physical laws. While unsettling, this scenario is possible however the WOMP doesn't show a Swiss cheese effect in space's BMR. Here the equation of the universe becomes R a b c d=R a b c +1/2 R g a b/R a b times 1/c2 instead of -1/2 R g a b because space-time is curved out rather than in.

CHAPTER 7

DEMONSTRATION THAT SPACETIME CURVATURE IS DESCRIBED BY THE SPIRAL FRACTAL FORMULA

The mathematical expression

$$N(pi/2)i \int \frac{du}{u}$$

DESCRIBES THE SPIRAL FRACTAL FORMULA WHICH WAS DESCRIBED BY ALBERT EINSTEIN IN 1912 to determines space-time curvature with any mass or masses due to the gravitational effect on space-time causes by mass.10 This is clearly seen when an expanding area or space-time that is also rotating is spiral when one takes a slice of space-time almost infinitely long. The expression du/u when integrated reveals In u which when integrated from –infinity to infinity reveals 0 or the quantum ground state >This was pointed out in this author's book "Megaphysics, A new Look at the Universe"(2003). In terms of quantum mechanics the reciprocal of In u is e to the u power. As a consequence -1/2 e –I to the n cot theta power relates directly to space-time curvature or the curvature metric R g a b in terms of tensors.

Dark Matter is basically undefined but is postulated as having baryonic particles and neutrinos. It is indirectly measured by gravitational perturbations, which would otherwise be difficult to explain. In this author's first book "Megaphysics, A New Look at the Universe" it was purported that the expansion of galaxies away from each other with the expansion of space-t which is increasing rather than decreasing with the unaccounted for mass which appears to be increasing indicates that the inertia from this unaccounted for mass should be decreasing the expansion of space-time and galaxies rather than increasing it. A logical conclusion is that the unaccounted for mass is either in other dimensions which cannot be perceived or directly measured or the unaccounted for mass is "measurement error".

In the equation -1/2e-I to the n cot theta indicates that the odd dimensions 1st, 3rd, 5th etc. are shadow dimensions containing matter that cannot be directly measured or perceived but can be indirectly measured. It also indicates that even dimensions 2nd, 4th,6th etc. can be directly measured and perceived. It is logical to presume that if dark matter exists it is sequestered in the odd dimensions and ordinary matter is sequestered in the even dimensions. It has been postulated that dark matter may act as cosmic glue holding together ordinary matter and possibly anti-matter; although anti-matter seems to lack cohesion because Dark Energy has the anti-gravitational effect of dark matter curves space-time with reciprocal curvature to the mass of ordinary matter. Dark Energy carries dissociated antiparticles out in expansion of space-time. It makes logical sense that "cosmic glue" would be in odd dimensions when ordinary matter and dissociated antimatter would be in even dimensions.

Space-time curvature is extreme at the event horizon of black holes and takes on a collapsing vortex leading into the black hole. This illustrates the strong force involved with fission and fusion and involving quarks in the nuclei of atoms and how they relate to space-time regarding collapsing mass.

In the case of electromagnetism one is dealing with negligible mass in electromagnetic fields as radiation is composed of photons, which have a mass of approximately 3x10-27 eV/photon and electrons whose mass, is 9x10-28 grams. This mass of a photon is experimentally derived from static electromagnetic field studies. As mass curves space-time and negligible mass curves space-time negligibly the effect of electromagnetic fields as a force on space-time is negligible based on the action formula .In it only during a "Big Crunch" or the event horizon of a black hole where radiation would be concentrated that the spiral nature of space-time is noted. Also as all measuring devices are part of what's being measured a Skew Symmetry following Heisenberg's Uncertainty Principle would show asymptotic flatness as space-time curvature only shows spiral components at extreme concentrations of mass such as black holes, "the Big Bang" with a swirl at the very inception of the blast at Planck's Time or during the ending phases of a "Big Crunch". A "Big Crunch" phenomenon is similar to a flush effect with the Coriolis effect.

CHAPTER 8

HOW SHADOW DIMENSIONS EXPLAIN DARK MATTER

AS PREVIOUSLY MENTIONED DARK MATTER CANNOT BE DIRECTLY MEASURED OR PERCEIVED. In odd dimensions where extreme mass has been indirectly detected and measured dark matter could easily be situated. Other phenomena could also be situated in those "shadow dimensions" which cannot be directly measured or perceived however paranormal phenomena albeit explainable will not be discussed in this book.

In the original quantum mechanics equation Planck'sMassXc2xthe Hamiltonian Operator with respect to n eigenstates of energy from a eigenstates of the probability of point r with respect to time times the La Placean Operator with respect to n dimensions=-1/2e-I to the power of n cotangent theta where n=the number of dimensions explains dark matter in the odd dimensions and ordinary matter and anti-matter in the even dimensions. It is still subject to Kurt Godel's axiom of incompleteness, which states that every set is a subset of another, set that while incorporating everything is still incomplete. This axiom replaced Formalism as purported by Hilbert who derived "Hilbert space" in quantum mechanics. Proving that n dimensions approaches infinity and completing the Unified Field Theory must go to Zeno's Paradox 11already mentioned which states that two objects halving the distance between them continuously down toward an infinitely small distance will never and can never touch. When one things of one second of arc or $1/3600^{th}$ of a degree this second of arc can be subdivided an infinite number of times indicating an infinite number of planes (an area between two points). These infinite planes are in motion though as previously mentioned in Chapter 6,and move with space-time in an accelerating expansion with a rotational vector, which is decelerating. According to the osculating plane these infinite dimensions twist turn and curve through the 26 dimensions defined by string theory and compactified (curled out and shrunk to string size) as well as every other dimension imaginable therefore incorporating all dimensions which are string sized and those which are more microscopic than Planck's Length (10-33cm). Space-time curvature is spiraled but incorporates all the mass that curves space-time including anti-matter, dark matter and ordinary matter and therefore traverses an infinite number of dimensions due to the osculating (twisting and turning) plane with respect to other planes. Will scientists ever be able to establish a boundary for space at Planck's Length (the size of a string)? With technology approaching what has been coined the Omega Point 12 where everything that is learnable has been learned it is not impossible that a civilization somewhere will or has determined that space exists at points smaller than Planck's Length. The Omega Point was mentioned in a book written by Dr. Frank L. Tipler who also authored The Anthropic Cosmological Principle. Based on this The Hamiltonian Operator showing quantum fluctuations over n dimensions can approach infinity but will not reach it due to Gödel's Axiom of Incompleteness.13

Considering the multiverse as postulated by other physicists including Lisa Randall in page 60 of her book "Warped Passages" 14 which states two particles may be too far apart to communicate with each other and have different manifolds or universes with their own laws of physics only sharing gravity with ours membranes or branes can twist or turn into an almost infinite number of configurations in a multiverse that may be huge . Note also that Alan Guth's Inflation hypothesis does not rule out multiverses in his Swiss cheese effect of our universe.

Chapter 9

THE ANTHROPIC PRINCIPLE AND Its RELATION TO THE MULTIVERSE

As mentioned in chapter 8 the Anthropic Principle was mentioned in a book written by John D. Barrow and Frank L. Tipler "The Anthropic Cosmological Principle". The Anthropic Principle states that "we are here and exist" and any physics must include a universe or space-time manifold with the right conditions for life and intelligent life to exist on this planet. Life is based on carbon on the third planet revolving around the star solaris or the sun, which exists in a galaxy, which has been, coined "The Milky Way". This life requires oxygen to exist and is in plentiful amount 21% in its atmosphere. Radiation that is toxic from the star solaris is filtered by an ozone layer in the stratosphere and ionized in the ionosphere. Intelligent life evolved as per the theory of evolution authored by Charles Darwin and intelligent life exists if any only if the observer of the phenomena exists and is not reproduced in the brain of the observer.

The multi-verse would include a universe with a "Milky Way" galaxy so it is consistent with the Anthropic Principle. A near infinite number of dimensions on a submicroscopic level is also consistent with the Anthropic Principle. Shadow dimensions and dark matter are consistent with the Anthropic Principle. A Big Bang and Big Crunch are consistent with the Anthropic Principle and curved space-time in a spiral configuration is also consistent with the Anthropic Principle as everything in spiral space-time would experience asymptotic flatness (no curvature as per the measuring device which is part of what's being measured. Two dimensional strings are not totally consistent with the Anthropic Principle if there isn't a third dimension that is smaller than Planck's Length (which there is and must be).Certainly M theory which is a composite of the string theories Heterotic 8x8,string theory type I string theory type II ,type II a and SO(32) string theories.14 Heterotic strings share two dimensions in one ;one rotating clockwise with a rotating orbifold or Calabi Yau Manifold and the other rotating counterclockwise. Notice that all planes can osculate or rotate they can define an infinite number of dimensions for each subdivided space in each second of arc. Conclusion strings and flat matter are not actually two-dimensional although the 2 dimensional equations appear to work with them. The other dimensions and membranes are so small that there part of the equations are not important. Duality is observing the same phenomenon from different vantage points. This is describing a horse from five different positions around the horse as described by a blind man. The descriptions are different but they explain the same horse. The five string theories which composed M(Membrane)Theory are all dual to each other but describe the same theory M Theory.

Two planes, which intersect, define a dimension. When two infinitely small slices of a second of arc and are osculating as in a moving frame of space-time it is a moving frame or moving triad. This triad is a mutually orthogonal triplet of unit vectors A,B and C it constitutes a right handed component of the basis elements for E3 where it moves continuously along C where I C is a plane that is osculating with reference to unit vectors A,B and C. The triad which changes continuously along C is the moving triad A,B and C whereby T and N which are orthonormal to vectors A ,B and C the planes T and N are osculating. An orthonormal vector is 90 degrees to the target vectors A,B and C. In terms of arc-length paramterization T=r./||r|| N=e times (r.r.)r..-(r.r..)r. /||r.||||r.xr.|| and B=e(r.xrr./||r.xr..|| Where e is epsilon not 2.71828 and the choice of sign of epsilon is + or – depending on the choice of N as aC 1 vectorC1 is called the principal normal vector which is orthonormal to th unit tangent vector if a vector lies in the plane of a curve for any non-straight planar portion of the curve. The unit tangent vector T=r'(dx/ds,dy.ds,dz/ds)15.Note ds/dt=||r.||footnote: Tensor Calculus David Kaye p.129. The reason the frame is right handed is because it constitutes right handed or clockwise curve for spiral space-time. When a left handed counterclockwise slice of space-time is osculating in a mathematically chiral manner then these two osculating planes intersect forming a dimension and as there are an infinite number of slices in each second of arc there are an infinite number of osculating dimensions. Calabi Yau manifolds are surfaces, which in terms of osculating planes twist and turn into the most stable configuration whereby they are acted upon by closed bosonic strings and their associated masses causing the curvature metric of space-time in motion with regard to a moving frame. According to string theory open strings have a discontinuity or free end, which can move in an almost infinite number of frames and closed strings are continuous but can twist turn flip or move in an almost infinite number of ways. Chiral osculating planes show an infinite number of intersections with an infinite number of subdivisions of space-time and orthonormal vectors of space with unit tangent vectors in expansion and rotation with right and left chiral components.

CHAPTER 10

STRING THEORY, M THEORY AND THE 0,0 POINT

String Theory was initially purported in the 1980's when it was discovered mathematically that nature follows harmonics of musical notes. These harmonics were incorporated into two dimensional energy components of matter called strings. There were two varieties of strings closed and open strings. Closed strings were continuous and formed loops, double torus, torus configurations, triangles, rectangles and a myriad of other configurations based on the energy of the string with regard to other strings in space-time. The unit of space-time was coined the orbifold; a geodesic of space-time which is a surface or manifold that twists and turns in myriad configurations as closed strings do. Calabi Yau manifolds are derived configurations of orbifolds and are generally Planck's Length 10-33 com. Open strings WERE COINED BY DR.ROGER PENROSE AS TWISTERS ANOTHER DESCRIPTION OF OPEN AND CLOSED STRINGS. Twister theory incorporates imaginary numbers with dimensions to indicate matter in a shadow universe, similar but mathematically different from this author's -1/2 e-I to the n cotan theta power incorporated in space-time with regards to the Quantum Mechanics equation 16. Art Planck's Length 10-33 space-time is curved in around itself like min-black holes giving it infinite curvature and transforming it into a quantum foam from continuous space-time according to quantum theory while Einstein purported that space-time was continuous down to an infinite size .ALSO,EINSTEIN STATED THAT WHAT WAS WHAT APPEARRED TO BE THE FORCE OF GRAVITY WAS IN FACT SPACE-TIME CURVATURE DUE TO MASS.P.18 Superstrings and the Search for the Theory of Everything.17 Gravity is an effect not a force and the effect is caused by mass curving space-time according to Einstein. This is why R g a b is the space-time curvature metric, which describes the EFFECT of gravity on space-time R causing curvature. String frequencies are based on the note frequency f=1/2L(T/m)1/2 where L is the string length T is the tension and m is the mass of the string. Decrease the length of a guitar string the frequency increases. Twist the string or guitar peg and the frequency increases. Place a finger at the midpoint of the string and the frequency splits in two. These concepts are the basis of string theory or as Pythagoras stated "the music of the spheres'" which was first thought of by the Greeks. Open strings join and split and closed loop strings mimic spin 2 vector bosons, which are quantized units of gravity which is the curvature of space-time. These particles are carrier particles which curve space-time for each any every target mass. In other words, spin 2 vector bosons actually curve the space-time manifold or surface and are instrumental mass components that do the curvature. Again this is the effect of gravity rather than it being a force again according to Einstein. Hadrons were explained which are strongly interacting elementary particles. These spin 2 vector bosons appear as closed loop strings and as a basic building block of matter spin 2 vector bosons are a basic block of matter which cause them to pervade all mass everywhere and curve space-time as that mass. One could get into the Higgs Boson which was coined "The God Particle" which would bear out the boson as being the fundamental building block of matter and energy all pervading. As previously mentioned the self dual hybrid of the five string theories type I, type ii, type IIa, Heterotic 8x8 and SO(32) is M Theory which is actually short for Matrix Theory although Membrane Theory has been used at times also. M THEORY CONTAINS SUPERGRAVITY IN THE 11th dimension (D=11)in the low energy limit and reduces to the type IIa string theory when compactified on a sphere. The infinite momentum limit of the D-0Brane may relate to the U(N)super-Yang Mills Theory. The D-0Brane relates to the 10 which broke up or cleaved into a dimensional state which has a limit of N approaching infinity for 10 dimensional 0-Branes.[18] Based on this idea there were an infinite number of dimensions in the vacuum state pre Big Bang and 0-Branes were raw building blocks of everything which were cleaved or broken into a six dimensional component and four dimensional component where the six dimensional component relates to the Calabi Yau Manifold and the four dimensional component relates to standard space-time resulting in the 10 compactified dimensions of Type IIa String Theory. This idea is consistent with a conservation of dimensions. Infinite momentum (p) relates to the Unified Energy with super gravity in the 11th dimension as described by Yang Mills U(N). To be compactified onto a circle type IIa String Theory would indicate a world sheet, which is two dimensional but spherical or circular in two dimensions. Closed strings would fit better in a world sheet which is compactified to a circle and this may eliminate problems with duality as the five string theories would curl up the world sheet to a point with infinite space-time curvature or a circle of quantum foam when the environment is below Planck's Length. In this author's previous book "Megaphysics, A New Look at the Universe" the term 0,0 point was coined by this author. The 0,0 point was in essence the point of the Big Bang or Swirl as there was a maximum rotational components from infinite curvature space-time in a point to asymptotically flat or curved space-time as with Inflation.

THERE WERE EQUAL AMOUNTS OF MATTER AND ANTIMATTER IN THE QUANTUM BUBBLE WHICH EXPLODED IN A 360 degree orb blast with a 180 degree trajectory where Riemannian 4 space for antimatter was given the tensor R jikl where kl is positive space-time.R jikl=R jiRkl times the partial derivative of x k power/x -1 times the partial derivative of l/with regard to x -1 for the covariant tensor and for the contravariant tensor R ji as subscript and kl as superscript=R jiR kl partial of x -1 with regard to x ktimes partial of x-1 with regard to x l.Using Riemannian 3 spaceR3(l,j,k) R ji=-e jiand Gravity for antimatter was R jikl-R ji with kl as superscript=-e ji= g ji/8(pi)G.R jikl-R ji as subscript and kl as superscript with R jikl as the covariant tensor for space-time and R ji with kl as superscript was the contravariant tensor for space-time   R jikl-Rji with kl as superscript=-8(pi)G e ji=g jiwhere -8(pi)G e ji is the formula for antigravity between antiparticles with antiparticle repulsion that triggered "The Big Bang" causing Dark Energy to push out with inflating space-time which is still expanding the galaxies outward with a progressively lesser rotational vector and increased expansion. The 0,0 point was where the extreme antigravity between all antiparticles, which lack cohesion due to its repulsion with other antiparticles, exploded with space-time expanding at over "c" the boundary by which no matter can exceed. There should and must be a black hole too far out for the light to ever reach us around the earliest galaxies where the 0,0 point exists; however with an isotropic universe a center is not easily found due to observational symmetry. In an anisotropic universe there may eventually be technology which will find this black hole if by then it hasn't dried up or evaporated however with a 6.75x10 34 erg blast the first second this black hole would have to be so massive that it would probably exceed the size of multiple galaxies and still be swirling inward with spiral space-time making it easier to detect. Note that the cosmologic constant^ is a very small number near 0 but reflects anti-gravitational effects of Dark Energy. This seems paradoxical but Dark Energy is carried with inflating and expanding space-time at velocities greater than the speed of light boundary. As the speed of light boundary was breeched "vacuum energy "may have carried the anti-gravitational Dark Energy outward in the expansion. Physicists are trying to seek a larger magnitude for Dark Energy as it should have been 6.75x10 34 erg during the first second after Planck's Time.

CHAPTER 11

SPOOKY ACTION AT A DISTANCE

Albert Einstein first discovered Spooky Action at a Distance when he discovered that when observed electrons were not in a particular position that they were supposed to be. Since then this was also found with photons and are referred to as photon pairing. The presence of an observer in an observational system affected the absolute location of electrons and there appeared to be an exchange of information between electrons and photons which were a huge distance apart from each other The probability distribution of electron pairs acting as an electron gas showed an electron field of

$$|eE| > E| > \frac{\pi m 2}{\ln 2} here\ \hbar =$$

$Planck's Constant\sqrt{\ } E - Ec\ has\ Ec\ as\ the\ conduction\ band\ and\ E > Ec\ Ec\ is\ the\ conduction\ band\ between\ which\ density\ of\ states\ r = 0\ The\ expression.$

(SOURCE: Wikipedia for electron cloud probability densities.)

The action of a Metric g from the mass of an electron with a carrier density of Ev as the state of matter acting on the electron gas has a probability density of 100 percent with regards to the localization of that electron in a gas of infinite radius.19 Therefore electron gasses vast distances away exchange information via Spooky Action at a Distance as the outer limits of the gas are an infinite distance from the observer and the observer can displace this via the Heisenberg Uncertainty Principle. Photon pairing is more difficult to document although the wavelength of a photon is described by the DeBroglie Equation h/mc where the wavelength can be demonstrated with respect to the energy of that photon.

SPOOKY ACTION AT A DISTANCE IS STILL DIFFICULT TO DOCUMENT IN TERMS OF MATH AND THE SPIN DIFFERENTIALS OF ELECTRONS IN AN ELECTRON CLOUD OVER LARGE DISTANCES WHICH CHANGE WITH THE

OBSERVER IS ALSO DIFFICULT TO DOCUMENT. Electron density is also related to the Boltzman Equation where an electron mass and the probability of a generating function penetrate a sphere of diameter infinity such that the diameter leads to an symptotic divergence of probability with a continuity limit for any field strength. Normalization of electron in the volume of a sphere can be summated in a system where the asymptotic behavior of the probability acts upon any particle masses with carrier densities. The FermiDirac Probability Function is calculated with the Boltzman Equation for specific states of an electron gas. Where

$$n = carrier\ density \int\ 8\pi\sqrt{\ }\ 2h\frac{3me3}{2}$$

SPACETIME=SPACE/MASS(1/c2) or spacetime=spacexk/mass where k=1/c2.

This translates in Relativity to R a b c d=R a b c -1/2R g a b for black holes the Big Bang and the Big Crunch. For ordinary expanding and rotating space-time R a b c d=R a b c+1/2 R g a b/R a b all multiplied by 1/c2 where c= speed of light boundary R a b=RicciTensor representing inertia R g ab representing the space-time curvature metric acting on R a b c flat Mintowski space to give Riemannian 4 space or Lorenzian curved space-time. In Quantum Mechnaics mass is described by Planck's Mass (h c/8 pi G)1/2{H al(r,t)|2d nr Laplacean operator for n dimensions=|(x,t)|2d n r Laplacean operator in n diemsnions(-1/2 e –I to the n cot theta power)where |(x,t)| is the probability of point x or the wave function of point x at time t in n dimensional space. The Hamiltonian operator H for a eigenstates of energy where a goes to n eigenstates of energy represents weak perturbations or quantum fluctuations of the

wave function $\psi(x,t)$

in n dimsnsional space and the LaPlacean Operator

∇n in n dimensional space represents the nth derivative with respect to x.. Riemannian 3 space

Mintkowski 3 space is represented by

$$|(r,t)|2 \, \nabla n \, d \, n \, r$$

which when multiplied by -1/2 e –I to the cn cot theta power causes the -1/2 e -1 n cot theta power to vanish as e –infinity power=1/e infinity power =0.Conversely -1/2 e I n cot theta power is infinity on the right and the Hamiltonian operator for a approaching n eigenstates of the probability of the wave function of (r,T) with the LaPlacean Operator over n dimensions times Planck's Mass which has a reciprocal of (8 pi G/hc)1/2 or

$$hc/8\pi G) -$$
$\frac{1}{2}(c2)$ approaches zero on the left side as the reciprocal of the Grand Unification Energy times

Planck'sMassxc2=10 19 Giga electron volts or 10 28 Giga electron volts in the denominator so the numerator doesn't matter and the quotient approaches 0. 0=0 as the right side is the multiplicand of -1/2e-I to the n cot theta power and -1/2e-I to the n cot theta power is e – infinity power =0. So again spacetime=space/mass (1/c2)

AS BOTH EXPRESSIONS ARE EQUAL TO THE SAME THING SPACETIME=k9SPACE/MASS WHERE K=1/C2 they are equal to each other BY THE TRANSITIVITY POSTULATE. THEREFORE REGARDLESS OF THE NUMBER OF DIMENSIONS QUANTUM MECHANICS AND RELATIVITY ARE UNIFIED BY THE EQUATION OF THE UNIVERSE SPACETIME=SPACE/MASS TIMES THE RECIPROCAL OF THE SPEED OF LIGHT.

The difference between 10 28 in the denominator on the left side of the Quantum Mechanics equation and infinity in the denominator of the right side of the equation which makes n/10 28 th power approximately equals 0=0 but the axiom of incompleteness still needs to be bridged by the Hamiltonian Operator of a eigenstates of energy in n dimensional state. So as a consequence c2(hc/8 pi G)1/2+H as a goes to n eigenstates of energy of the probability of point x at time t in an n dimensional state by adding the Hamiltonian Operator to the Grand Unification Energy of Planck'sMassxc2. This bridging causes the expression to go to infinity in the denominator of both the left and right side of the equation making 0=0 exactly, SO THE FINAL QUANTUM MECHANICS EXPRESSION OF SPACETIME=SPACE/MASS(1/C2)IS AS FOLLOWS THE WAVE FUNCTION OF(r,t)d n r times the LaPlacean Operator in n dimensions/c2(hc/8 pi G)1/2+H as a goes to n eigenstates of energy of the probability of (r,t) point r at time t in n dimensional space d n r LaPlacean Operator to the nth derivative=the wave function of (r,t)d n r times the LaPlacean Operator to the nth derivative-1/2 e –I to the n cot theta power where n=number of dimensions formulating spacetime as Riemannian or Mintowski flat 3 space with spacetime curvature metric acting on it.

$$\psi(r,t)d\,\frac{nr\nabla n}{c2\sqrt{}}\hbar\frac{c}{8\pi G}+\mathcal{H}a \to n(\quad |r,t)|d\,n\,r\,\nabla n = \psi(r,t)d\,n\,r\,\nabla n - \frac{1}{2e} - i\,n\cot\theta$$

This powerful expression where I is to the n cotangent theta power and theta is the trajectory of space-time or 180 degrees or pi radians at the Big Bang reveals 0=0 exactly for the Quantum Mechanics expression of space-time=space/mass (1/c2) and equals R a b c d=R a b c-1/2 R g a b/R a b all multiplied by 1/c2 in Relativity unifying them without Gödel's Axiom of Incompleteness. Note i=the square root of -1.

"A TRUE SCIENTIST ASKS QUESTIONS OF HIMSELF THROUGH THE SCIENTIFIC METHOD AND ANSWERS THEM TO THE WORLD"

GLOSSARY

Abelian: equations having a coefficient or variety in a specific groupe.g.,algebraic number fields, tensors of the same degree or cohominy group

Anisotropic: not isotropic, lacking observational symmetry

Anti-symmetric: tensors or vectors that are equal but opposite and can therefore cancel or partially cancel

Asymptotic: that which approaches a level or boundary but never reaches it.I,e.Asymptotic flatness appears without curvature but doesn't reach it

Bianchi's Identity: The identity of groups of 4 Riemannian Space that is anti-symmetric and Abelian and cancels each other out being equal but opposite

 "The Big Bang" : A theory proposed describing a Freidman type I open flat expanding universe which is homogeneous and isotropic (ex nihilo)would be out of nothing or spacelessness. Backed up by the background microwave radiation with the WOMP

Big Swirl: A Big Bang with a pressively decreasing rotational vector from an infinite curvature point of space-time to asymptotic flatness

Black Hole: collapsed matter from a neutron star, galaxy or universe with extreme density extreme curvature of space-time at the center(gravity)and inability to reflect light due to the extreme gravity of the spiral space-time at the event horizon(that area where the black hole is located)Entropy approaches 0.29 or 0

Calabi Yau Manifold: a surface which represents a relative isotropic portion of space-time with a puckering to accommodate multiple dimensions considered a twisted variant of the orb fold

Chaos: absolute disorder

Chiral: a mirror image or absolute symmetry

Closed string: a two dimensional building block of matter and energy. Movements consistent with observational behavior in the space-time continuum without breaking the continuous string. Strings are often considered as energy.

Compactified: A term used by Kaluza and Klein to call a curling up of dimensions mathematically as the size approaches zero.

Conformal space; when every point in space relative to every other point in space maintains its relative position regardless of what the space is doing

Dark Matter: an indirectly measured mass causing perturbations in gravity postulated as possibly acting as cosmic glue containing possibly baryonic particles and neutrino. Not directly observed or directly measured otherwise no specific definition

Entropy: degree of disorder

Event Horizon: area where a black hole is perceived to be by measurements

Ex nihilo: from nothing or a lack of dimensions

M(MATRIX)THEORY: A fusion of the five string theories which are sharing in duality to each other. Based on type IIa string theory being compactified onto a circle with membranes incorporated into matter. A congromerate vantage point description of matter based on stress density or concentration such that uniformity is met where time, matter and interwoven stress energies as described as two dimensional flat strings with integrated motions moving in a predictable or unpredictable motion or motions in the space-time continuum

Isotropic: observational symmetry, relatively non self-annihilating

Geodesic: a unit of space-time

Membranes: a description of matter in terms of stress energy density as described in the number of states with regard to dimensions

Mintkowski Space: four dimensional space-time with regard to length, width ,height and time; doesn't necessarily interact with Riemannian Surfaces

N: the number of dimensions in n dimensional space where n doesn't equal 0

Open string: a two dimensional building block of matter with movements in a multidimensional plane consistent with observations. Open strings have the capacity to break and reform and can be discontinuous and are also postulated a strings of energy

Orbifold: space-time manifold or surface, which rotates on an axis which is asymptotic appearance of a cone but is discontinuous.

Relativity: the behavior of matter and energy with regard to other matter, energy and space where matter has a different vantage point from other matter, energy or space including stress, energy, time and mass changes with regards to velocities of each component of matter in conformal space-time

Ricci Tensor: that tensor which represents inertial mass or inertia the resistance against push or pull by an unbalanced force or space-time curvature metric

Riemann forces: inclusive of all strong and weak forces strong and weak perturbations

Riemannian Space: Mintkowski space with Riemann surfaces in which matter and its energy density can interact

Scalar: the magnitude involved in a tensor without necessarily having regard to vector motion

Six dimensional string manifold: curled up closed strings in configuration according to Kaluza and Klein which is 10-33 cm and may be composed of Calabi Yau Manifolds.

Space-time manifold: Matrix over a Riemann surface in which energy density of matter exists as per Poisson's Equation: The derivative operator of a dual vector field=4 pi (rho)where rho is the energy density of matter. Note Gauss' Law was based on the Poisson Equation where the dual vector field is electromagnetic (E)

Stress: energy density with components of the strong force (quarks in the nucleus of any atom) weak forces (radioactive decay), electromagnetic force, inertia and gravity (which is an effect of space-time curvature due to mass) Strong and weak perturbatory forces interacting with matter in a certain way

Tensor: a force vector with subcomponents interacting with or without other vectors in either a moving or stationary frame where the magnitude only is called the scalar trace

Throat: a region of space-time, which connects with another region of space-time and is not discontinuous

Weyl's Tensor: a pull or conglomerate pull in a conformal manner. The pull represents space-time curvature in a conformal surface or surfaces due to inertia or inertial mass (Ricci Tensor)gravity

X or r: any observable in the space-time continuum (without beginning or ending)

Yang Mills Notation: diagrammatically configuring the different forces of nature which may or may not interact

Impossible: that which is beyond the comprehension of man (kind)

Science: a process to obtain knowledge

FOOTNOTES AND BIBLIOGRAPHY

FOOTNOTES

1:Peat,F. David.Superstring and the Search for the Theory of Everything. Yang Mills Forces p.114

2.Kaku,Michio.Strings,Conformal Fields and M theory.Ising Model p.176-78

3:Maxwell's Equations Wikipedia.en.wikipedia.org/wiki/Maxwell%27s -equations

4:Metric tensor(General Relativity)Wikipediaa and Spacetime.en.m.wikipedia.org.spiral Space-time Einstein 1912 Fractal Time.p.108-109 Braden,Gregg 2009 Library of Congress

5.Peat,F.David.Superstirngs and the Search for the Theory of Everything.p.106-107

6.Kaku,Michio.Quantum Field Theory.p.648

 7.Peat,F.David.Superstrings and the Search for the Theory of Everything.p.156-161

8.Kay,David C.Tensor Calculusp.129 Osculating Plane

9.Kaku,Michio.Strings,Conformal Fields and M theory.p.505

10.Peebles,P.J.E.Principles of Physical Cosmology.p.365 and 392

11.Zeno's paradox:The Math Forum at Drexel University

12:Tipler,Frank j.The Physics of Immortality

13.Godel,Kurt.Godel's Incompleteness Theorems en.m.wikipedia.org

14.Randall,Lisa. Warped Passages p.60

14:Green,Brian.The Elegant Universe.

and

14.Wick,Mitchell Albert .Megaphysics,A New Look at the Universe.Introduction.

15:Kay,David.C.Tensor Calculus.p.129 ..16:ibid item#7p.237-242 17:ibid item 16p.18

18:Kaku,Michio.Strings,Conformal Fields and M Theory.p.464-468

19:Wikipedia.Electronen.wikipedia.org/wiki/Electron

20:Spooky Action at a Distance.Quantum EntanglementWikipedia. Or en.wikipedia.org/wiki/Quantum entanglement

BIBLIOGRAPHY

Barrero,John D.The Anthropic Cosmological Principle.Oxford England.Oxford Press.1986

Brade,Gregg.fractal Time 2009 Library of Congress.

Greene,Brian.The Elegant Universe.NewYork.Vintage Books editor Random Press.1999

Hawking,Steven and Penrose,Roger.The Nature of Space and TimePrinceton,N>J>Princeton Science Library 1996

Kaku,Michio.Quantum Field Theory.A Modern Introduction.Oxford university Press.1993

Kaku,Michio.Strings,Conformal Fiels,and M theory 2nd edition.Springer Press.2000.

Kay,David C.Tensor Calulus Schaum's Outline Series. N.Y.McGraw Hill 1998.

Peat,F.David.Superstrings and the Search for the Theory of Everything.Chicago.Contemporary Books 1998

Peebles,P.J.E.Principles of Physical Cosmology.Princeton Series in Physics.Princeton University Press 1993

Wald,Robert m.General Relativity.Chicago, Illinose.University of Chicago Press 1984

Wikipedia: online encyclopedia.

Randall, .Lisa .Warped Passages HarperCollins Publishers.N.Y.2005

Tipler ,Frank J. Physics and Immortality. Anchor Books division of Random House.1993

APPENDIX

BRANES;A STRING IS A one-BRANE WHICH COUPLES TO A BACKGROUND SECOND DEGREE TENSOR.Zero-branes are ten dimensional building blocks for space in the pre-Big Bang epoch. The second degree tensor is purported of negligible mass as indicated by the ZERO-BRANE.THE SOURCE OF THE BACKGROUND SECOND DEGREE TENSOR IS R uv where the integral of D to the d power of x where x is the string or one-brane in D dimensions applies to R u v g u v where g u v is the metric acting on R u v for the zero-brane with respect to x which is the one-brane. R u v is the second degree tensor upon which the metric g u v acts. In four dimensions a monopole is dual to two electrons acting on a zero-brane. In 10 dimensions a string is analogous to a five-brane based on p-brane potentials. This involves dual fields such as a tensor R=R* from Ra1…a n=R8b1…b n.p-branes are encircled by a hypersphere which relates to M theory being compactified (curled up)by a circle for typeIIa strings.The charge of a p-brane is based on Q=

Are associated with a tensor of the pth rank R a1…a p and electric and magnetic charges can be associated with p-branes with superalgebra.

Dzero-branes represent the vacuum state.ALTHOUGH INDICATED AS ten dimensional building blocks of space they actually are zero-dimensional. ONE-BRANES REPRESENT STRINGS WHICH ARE TWO DIMENSIONAL OR POSSIBLY ONE DIMENSIONAL. If all the dimensions in a system or universe are conserved such that the total number of dimensions are constant;then zero branes would have to be ten dimensional in the vacuum state. Six dimensions for CalabiYau Manifolds and four dimensions of space-time. As the "c" boundary is approached infinite mass with reducing length and width occur when length becoming infinite.In this case width and height approach zero but do not reach it and become infinitely small curled up and compactiifed.In general although showing duality between different systems which are Abelian membranes are described by the forces involved with mass or energy associated with the membrane with reference of n-dimensional space where n dimensions would have n-1membranes or n-1 brane.

FLAT OR MINTKOWSKI SPACE IS DESCRIBED MATHEMATICALLY AS THE LINE ELEMENT OR ds2=dx2+dy2+dz2-c2dt2.FLAT SPACE-TIME IS SPACE-TIME WITHOUT ANY CURVATURE IN OTHER WORDS A VACUUM STATE HERE R g a b=0 which indicates that the space-time curvature metric=0 and therefore gravity =0 in the vacuum state.

CURVED SPACE-TIME IS GENERALLY DESCRIBED BY ds2=e-k|r| (dx2+dy2+dz2-c2dt2)+dr2 where r=space-time curvature metric described by tensor as R g a b.R g a b or r is determined by the inertial mass of the object doing the curving and the curving is performed by bosons and possibly gravitons or fermions.Spiral space-time has k=-i(the square root of-1)to the n cotangent theta power as suggested by Dr. Roger Penrose and proposed by this author.

MANIFOLDS

THE SIMPLEST MANIFOLDS ARE CARTESIAN SPACES WHERE A MANIFOLD STRUCTURE OR SURFACE IN TERMS OF TOPOLOGIES IS R to the d power with what's called an identity map Rd implies R d .The coordinate functions of this map are Cartesian coordinates. If coordinates are a I ;R d is the manifold of the standard Cartesian coordinates.a i=ax+ay+az and R to the d power is the tolological expression of the standard manifold or the Cartesian Coordinate system. If a manifold is imbedded in another manifold it is a submanifold. On a string basis submanifolds can be orbifolds or Calabi Yau manifolds which are submanifolds for spiral manifolds for asymptotically flat but curved space-time on a macrostatic surface which is expanding and simultaneous rotating as at black hole event horizon.

RIEMANNIAN CURVATURE A RIEMANNIAN SPACE IS THE SPACE COORDINIZED BY xi(power)with a fundamental form of the Riemannian Metric g I jdx I dx j where g=(g ij) obeys the metric tensor.g is of differentiability class C2(all second order partial derivatives of g I j exist and are continuous.g is symmetric g I j=g ji;g is nonsingular |g I j| doesn't equal 0.The differential form and distance from g isn't variant with regard to changes in coordinates.

R I j k l=g I ir(Rr superscript with jkl as a subscript where R jkl with ias a superscript is the Riemann tensor of the second kind. The Riemann Tensor of the first kind is R I j k l=Here

Rijkl=g I iRjkl is subscript and I as superscript in the diagonal metric te tensor calculations for the Riemann Metricgives six cases R one R 212 and 1 R 313 and 1 R 323 and 1 R 213 and 1 R 232 and 1 R 123 and 1 which proves with the partial derivatives of the Christoffel symbols of tensors according to the previous formulas give R I j k l=0 for all I j k and I indicating the summation of all Riemann forces and space is zero. The math of all these combinations is very difficult to reproduce by typing.

Alphabetical Index

3 CHAPTER THREE

THE FIRST EVENT

INITIALLY THERE WAS INFINITELY DILATED TIME(infinitely slow)and there were an infinite number of completely parallel planes. These planes were kept apart by the Cosmologic Constant which summed up to be Dark Energy. These were what is known in M theory as D-0-branes with fermions moving time in one direction and tachyons moving time in the opposite direction. The point of creation as when rogue tachyon or tachyons dropped from above the speed of light to the speed of light causing a time oscillation paradox with a spin magnetic moment forming monopoles and a centrifuge effect forming the string dimensions and macroscopic dimensions. This in turn formed the spacetime continuum which was spiral in configuration.

CHAPTER IX (9)

WHAT IS "THE EQUATION OF EVERYTHING" AND WHAT DOES IT TELL US?

The equation of everything as stressed in this author's previous books is based

on the inch long expression space-time=space/mass. As explained in the book

"Megaphysics II ;An Explanation of Nature" the expression of space-time in terms of

metric tensors which are vectors in multiple dimensions acting upon and being

acted upon by other tensors is the Lorenzian Tensor of the fourth degree described

as R abcd where R is the magnitude or scalar REGION IN TOPOLOGIC SPACE AND

abcd are the permutations of all the vector directions interacting upon and being

interacted by the space-time vector. This can be illustrated by a 4x4 matrix and

incorporates space-time identities such as Bianchi's Identity which show that many

of the vector components cancel each other out with regard to space-time still

incorporated into 256 permutations. R abcd is curved space-time and the

derivative of that tensor illustrated by th Christoffel Symbol Γ is acted upon all the

eigenstates (quantum states) of energy from the ground state energy level as

Riemann forces of nature.

described by the Cosmologic Constant (energy density of a vaccum) to the maximum

excited state or the Grand Unification Energy of 10^19 giga electron volts or 10^77

joules as describing the massive Higgs Field. To incorporate entanglement in these

disparate energy levels one must use $1/\sqrt{N}$ WHERE THE EIGENSTATES ARE

INTERMIXED OR INTERTWINED THROUGHOUT WITH ENTANGLEMENT AND N> or

equal to 2 which is synonymous with the "on " and "off switches of a quantum

computer entanglement.   Using shorthand

$1/\sqrt{N}$ $acts\ upon\ \Lambda\ 0|10^{77}joules\,T\ max$ |where T max is the maximum tensions

of each string   or

$4\pi\rho$ where $\rho = energy\ density\ of\ matter\ of\ the\ strings'\,tension.$

The above expressions act upon and are acted upon by space-time curving it inward

(gravity effect) or outward (anti-gravity effect) bringing one from a vacuum state

described by the Cosmologic Constant to the maximum energy state of 10^77 joules

or 10^19 giga-electron volts and described many tons of tension of on each open or

closed string . Poisson's equation utilized the dual vector field of string tension

Riemann forces of nature.

times the number of strings in the topologic region of space-time and the quantum

eigen-states of energy. The left side of the "Equation of Everything" equals the

vector sum of the Cosmologic Constant + or –(representing the positive and

negative Riemann Forces of Nature) R a b c which represents flat Euclidian Space +

or – ½ R g ab (representing anti-gravity as the + sign as anti-gravity curves space-

time outwardly flattening it and – as it curves space-time inwardly as time constricts

space toward the infinite curvature of a point or approaching zero diameter but

limited by Planck Length or 10^-33 cm which is the size of a string. This expression

reflects "space" and is divided by mass as described by the Ricci Tensor or I

$\hbar \rho\ abc\ and\ \hbar =$

$h \frac{}{2\pi} where\ 2\pi\ in\ the\ denominator\ of\ the\ denominator\ moves\ into\ the\ numerator$

and h is Planck's Constant or 6.63x10^34 joule-seconds .This is multiplied by the

energy density of matter

$\rho\ abc\ das\ tensors\ of\ the\ third\ degree\ acting\ upon\ flat\ eucidian\ space\ and\ for\ curved\ sp\quad pabc$

The makes the expression compactified to a circle where the circumference C is

Riemann forces of nature.   3

curved space-time over the total number of eigen-states of energy with

entanglement and the cosmologic constant is added to

2π Radius where the radii are the positive and negative.

This describes zero external field limit which approximates the limit where

fracture of the lattice of quantum dots may occur localizing spacelessness.

Basically space appears to be infinite in magnitude and direction but space-time

appears to be infinite in magnitude but finite netting zero in direction. These

observations may relate to times arrow moving forward or backward. At the speed

of light boundary "c" space-time approaches zero as time dilates toward infinity. At

the event horizon of any black hole space-time approaches zero as time dilates

toward infinity. Finally, the question of whether nothing exists or not depends on

whether space is infinite with an infinite sum of infinity with regard to magnitude

and direction or if space is 0(zero) with regard to direction while it may be infinite

with respect to magnitude which would include what this author calls "an infinite

space vacuum "or pseudo vacuum if energy exists in infinite space as compared to

absolute spacelessness.

Based on the Relativistic form of the Equation of Everything $\mathbb{R}\,a\,b\,c\,d=R\,a\,b\,c-1/2R$ $g\,a\,b/R\,a\,b$ as the mass reflected by $R\,a\,b$ goes up $R\,g\,ab$ the space-time curvature variant from the metric $g\,a\,b$ representing the effect of gravity on flat or Mintkowski Space $R\,a\,b\,c$ can curve space from an asymptotic flatness to an infinitely curved point at the event horizon of a black hole making the expression $0=0/R\,ab$ where $R\,a\,b$ represents the inertial mass . Based on this space-time can exist at an infinitely or near infinitely small size with infinite curvature. This still isn't spacelessness .It is the opinion of this author that spacelessness is a boundary or interface which requires a force which may equal or exceed a "Big Crunch "implosion to breech however as an asymptotic boundary it may not be breechable even with the implosion of a "Big Crunch" .The term impossible means "that which is beyond the comprehension of mankind" as quantum mechanics says every permutation that can occur does occur therefore spacelessness if and only if a permutation of space and space-time can and will occur.

The equation of a circle arc length(s)=rθ where r=radius if the arc length

approaches zero and the

angleθ *approaches zero they are AYSMPTOTIC FUNCTIONS* AND AS ONE

SECOND OF ARC CAN BE DIVIDED DOWN AN INFINITE NUMBER OF TIMES

WITHOUT REACHING ZERO THEN THE ARC LENGTH REPRESENTING THE

MAGNITUDE OF SPACE-TIME WOULD BE DIVISIBLE DOWN AN INFINITE NUMBER

OF TIMES AND NEVER REACH ZERO.THIS IS WHERE S=SPACE-TIME AND THETA IS

THE SPACE-TIME CURVATURE METRIC OF ANY AND ALL "r". This corollary

assumes the axiom that IIa string theory compactifies to be a circle   which is a

documented fact. Therefore the tensors of the fourth degree whose magnitudes are

additive whether positive or negative and whose directions are positive or negative

in a near infinite number of intersections of osculating planes then net space-time

can be zero but due to LOCALITY zero is virtually never reached or measured by

intelligent observers. When a measuring device is local and part of being measured

the observational viewpoint of zero space is beyond the comprehension of mankind

,however not necessarily beyond the observation of other intelligent observers which are not local. THIS IS IMPORTANT. The reason why nothing doesn't exist is

because based on the collective cognitive abilities and information gathered by a

hypothetical or real observer or observers is being the comprehension of that

hypothetical or real intelligent observer or observers .Mathematically the null set

has been derived and used many times by mathematicians and physicists and is

therefore considered a real finding in nature but mathematical computation and

comprehension of a phenomenon are two different and distinct things.

CHAPTER TWELVE

THE ANTHROPIC PRINCIPLE

The anthropic principle states one and only one thing. We are here. We exist.

Based on this idea there must be a specific combination of events to make possible

intelligent life . This series of events includes the formation of amino acids ,RNA

,DNA ,oxygen, water and an environment which is totally conducive with the

formation of intelligent life. This in essence is the sum total of the anthropic

principle but to amplify it there are ideas that without "intelligent observers" an event cannot be proven to have occurred. The question "When a tree falls in the forest and there's nobody around to hear it does it make a noise?" The answer isn't obvious. One would think using judgment that the answer is yes as the Physics of a tree falling with an atmosphere would create sound waves which follow the Doppler Effect .The fact that there's nobody around to hear it doesn't affect the formation of the Doppler Waves creating sound. It is true that there would be no body around to hear it or measure it with a detector but that doesn't mean the sound doesn't exist.

In Quantum Mechanics an intelligent observer affects measurements within an uncertainty range and there are no absolute measurements only ranges which is why measurements involve an expectation value<> and probabilities $||x||^2$ rather than absolute measurements. This is because the measuring device affects what is measured and LOCALITY EXISTS IN SPECIFIC STATES where local phenomena are affected by everything in that region including measuring devices. However while it's true that the sound made by a falling tree in a forest will and must be affected by an intelligent observer being present. the effect should be miniscule. ALSO IF AN INTELLIGENT OBSERVER HAPPENS TO BE DEAF OR DEAF AND BLIND AND IS

PRESENT IN THE FOREST WHEN THE TREE FALLS, A NOISE IS STILL MADE.

Therefore "common sense" says the answer must be yes ,but it is affected by the intelligent observer being present. This implies that all events in space-time which occurred before the formation of life or intelligent life did indeed occur ,as the converse is unlikely .If there was no intelligent observers during the first 12 billion years after the "Big Bang" how could the Milky Way" and the planet earth have been formed. Without the milky way and the planet earth being formed ,how could we exist as intelligent observers? The converse of this case is if other intelligent observers existed going back to the Big Bang. Who or what are the intelligent observers that existed during "The Big Bang"? The answers go into the purview of religion which is not the objective of this book although The Higgs Field and Higgs Boson have been postulated as "The God Particle" and the Higgs Boson was isolated in Cern ,Switzerland by the Hadron Collider. Prior to this , this author had some evidence that the non-isolated tachyon may have been the God Particle as it acted like negative mass rather than positive mass and appeared to go backwards in time where times arrow was still moving forward rather than backwards. Also the tachyon apparent broke other rules of physics such as exceeding "c". As the tachyon

apparently becomes a boson at v<c when it is no longer a tachyon, it could still be a

property of the massive Higgs Bosonic Field.

Is There Life Elsewhere?

The laws of probability say the answer to this question is undoubtedly "yes". Is

there intelligent life elsewhere? The answer to this question according to

probability is also" yes." Is there local intelligent life elsewhere which is not related

or branch to our life forms? The answer to that by probability is "no". It is likely that

on moons such as Io or distant newly detected moons in other solar systems there is

an environment to support life, but the likelihood that this life is either vegatable

,fungal ,bacterial or viral. Animal life is so complex in its intricacies that to have a

celestial body with animal life(even simple animal life) would have to be large

distances from the earth's detecting devices. It is certainly possible that there is

intelligent life in The Milky Way, however the frequency of this occurrence would be

so low that this author might be talking about fewer than five celestial bodies; while for bacterial or fungal life or vegetation the number of celestial bodies may be in the thousands. It is also possible that there are no celestial bodies with intelligent life outside the earth or perhaps only one in this vast galaxy. It may take many years to answer these questions if at all.

Was there ever life on Mars?

Missions to Mars include the Mariner series , Viking 1 and 2,Mars Observer ,Global Surveyer, Pathfinder ,Climate Orbiter ,Polar Lander ,Deep Space 2,and Phoenix. Also 2201 Mars Odyssey, Exploration Royers, Mars Express and Reconnaissance Orbiter ,Mars Science Laboratory and Maven. Mars may have water and in the past an atomosphere which means that sub terranium life may have in the past existed or may exist now under ice. Gathering information about "The Red Planet" takes years and may or may not give a direct rather than indirect answer. Bacterial, fungal or viral life is probably more likely than plant or animal life. Certainly to have photosynthesis it would be difficult to occur beneath the surface of the red planet and while it's possible that fossils may exist it still be many years to get the answers. Underground and bacteria or viruses it may exist or may not.

Moons such as Jupiter's Io were also been considered life friendly due to volcanism.

It is unlikely that our civilization will have concrete answers in the near future. If

one trust's one's senses and perception we are here. That is a fact. Do others exist

besides the individual that is perceiving the environment your brain says "yes' to

you but is the answer yes .We make assumptions about our environment and then

follow its rules. Would physics exist without life ? If there's life in this universe is the

universe alive? We have obvious answers to these questions   ,but are they correct?

These answers delve into philosophy .Can the reader prove he or she cohabits this

universe with others? By the assumption that the reader isn't the author the reader

can convince himself or herself that it's true. Physics SHOULD exist with or without

life but can't be proven as it takes life to prove it. If there's life in this universe and

the life is part of the universe then the universe is alive, at least in part. Do others

exist besides the reader? Probably because the reader's brain can prove it to the

reader. Is there more to this universe than what is measured and perceived? This is

unknown and may remain unknown as a cell cannot perceive a body it inhabits or if it is independent. We are   here can be amended to "I am here" or "you are here

"being the reader.

The Anthropic Principle puts constraints on the definition of life   ,intelligent or

not .It implies that life requires an environment conducive of life which can require

food(ingestion or photosynthesis),respiration ,have reproduction, locomotion(very

slight in plant life)excretion ,and can be born and die. It states that conditions that

form "us" require D.N.A. ,oxygen, ambient temperatures and water as well as a P.H.

which will not destroy important tissue .It deems it unlikely therefore that the entire

universe is alive Veneziano(1993)18. Although galaxies rotate and in dilated time

can act as dynamos producing energy ,black holes can act as an excretory function

such as Mallipigian Tubules and "The Big Bang" can act similar to fertilization. This

supposedly preposterous hypothesis would answer the question "What is the

meaning of life"besides perpertuating life would show that every microcosm of life

have a function in pertpetuating the whole ,although in terms of dilated time an

organism such as us could live only an instant in the scheme of things just as a

bacterium would live only a few days compared to a human being. Viruses perpetuate viruses ,bacteria perpetuate bacteria ,plants perpetuate plants and

animals perpetuate animals. The food cycle perpetuates life on earth and the

environment .Intelligent observers may perpetuate life on this planet by preventing

global catastrophes such as an asteroid strike or perpetuate distruction such as with

a cobalt bomb or H Bomb. How would life on the planet earth perpetuate a "living

unverse' in dilated time. The answer to this question is unknown and may never be

known to us;but that doesn't mean it automatically means "it can't".

Existentialism states that "life is meaningless".Is this true? I life has a function in
perpertuating life on a larger scale then the statement would be false but on a
smaller scale life does perpetuate life.or attempts to . We are trying to prove that
"we are not alone in this universe"and while we are closer than in the 1960's or
1970's we have not proven it yet as of this date in early 2015.
    Paranormal phenomena are used to determine that life exists after death and
while there is extensive anecdotal evidence of such life,it is difficult to prove
empirically with scientific measurement except for a drop in kvp and temperature
around "cold spots" in the environment where orbs exist and paranormal
phenomena have been reported.

CHAPTER THIRTEEN

QUANTUM MECHANICS

It is impossible to absolutely measure or localize anything measurable in space-
time. Everything measured must be in range based on the probability of any point
particle $|(r ,t)|^2$ and is described by the wave function$\psi(r, t)$. This is based on the
Heisenberg Uncertainty Principle and defines the Schrodinger Equation.
    Assuming the velocity of space-time is
πc or the product of π and c or the speed of light and that the sum total of all Riemann Forces

and energy is zero then the wave function of any point particle must travel at less
thanπc which is the speed of space —

time. Each point particle has a deBroglie wavelength which is $\lambda =$
$\hbar \dfrac{}{mc}$ *so there are wave properties to any matter with mass although photons have a resting m*

mass approaching zero while there is a measurable mass while photons are in motion. Before discussing this topic one must prove that space-time travels at pi(c).In a circle which is the compactified form of type IIa string theory the circumference describing space-time approaches infinity with reaching it and the diameter involving the sum total of all Riemann forces approach zero(0)with a domain of -

$\infty\, to\, \infty . c/d = \dfrac{\infty}{0}$ *with a domain of* $-$
$\infty\, to\, \infty$ *including the null set of* $0.$ *The denominator is the null set* $\{0\}$*where space* $-$
time is a open dinvergent set and 0 is a closed convergent
set. $\dfrac{\infty}{\infty} = $ *everything except* 0 *and* $\dfrac{0}{0}$ *is everything including* $0. \pi =$
$\dfrac{\frac{22}{7} and \frac{1}{\pi} is \frac{7}{22} so \infty}{0}$ *times* $\dfrac{0}{\infty}$ *is* $0.$ *Assuming that space* $-$ *time* $=$
circumference and diameter is mass and energy $=$
$R\, ab \left(\dfrac{1}{c2}\right)$ *of the metric* $g\, a\, b$*doing the curving on* $R\, a\, b\, c$ *which is Mintkowski Space or Rimenar*

Space-time.Again
θ *is the space* $-$ *time curvature variant* $R\, a\, b\, c\, -$
$\dfrac{1}{2} R\, g\, ab$ *where* $R\, g\, ab$ *is the curvature variant from metric* $g\, a\, b. \pi$*is the circumference of* $a \dfrac{c}{dia}$

Circle/diameter.s=r(theta) where s= space-time,r is the sum total of Riemann forces and theta is the space-time curvature metric .The circumference of a circle is without beginning or ending therefore converges to two pi radians or 360 degrees as an asymptotic function without reaching it. Therefore the absolute measurement of a circumference is is not possible but only an approximation of a convergent function to two pi radians or 360 degrees. Therefore the space-time continuum as the circumference being in the numerator/total Riemann forces in the denominator with +forces/-forces/matter and anti-matter causes curvature and reciprocal curvature of space-time between the +forces and – forces causing the swirl and anti swirl of The Big Bang" .Space-time travels at pi times the speed of light as it must do so to exceed the Riemann forces to "outrun them ".Again the circumference of a circle defines a continuum as it is without a beginning or ending and converges to 2 pi radians therefore going to infinity .The diameter isn't bounded as the matter and energy can never reach the circumference as it can't exceed "c" or the speed of light boundary while space-time travels at v>c by pi .Without boundaries for matter and energy the diameter can travel outward a near infinite distance without reaching infinity .As a result the equation space-time/mass can never actually reach pi or 22/7 which is why pi goes on to infinity without reaching a definite decimal termination 3.1415928....If the diameter of Riemann forces was bounded space-time would be slowed down to v<c or v=c in a "Big Crunch "and everything would be

compactified to pi radians or 180 degrees which is the trajectory of "The Big Bang
".This is when the sum of Riemann Forces=1 so the circumference/diameter
becomes 22/7 divided by 1.Aagin above v=c the circumference travels at v=pi©
while the diameter travels at v<c including light and photons which do not travel at
c but just below it as "c" is a boundary to everything except space-time and tachyons
.In this universe the Riemann forces must converge in a "Big Crunch "and divergent
space-time will also converge from infinity to 22/7 or pi radians while the
denominator or Riemann forces converge to 1 yielding pi radians at the central core
or the quantum bubble of the "Big Crunch" Times arrow must also point backwards
in this case as infinite space-time is converging to pi radians. Finally   ,you can not
absolutely measure the circumference of a circle (as mentioned previously)as it has
no beginning or end so the number of rotations of the circumference must be n
where n approaches infinity or infinite number of turns of 2 (pi)radians each as well
as an infinite number of subdivisions. Therefore infinity/0 approaches 22/7 in a
bounded convergent state which is the circumference of a circle/diameter that is
where the domain goes from –infinity to +infinity to
$0<\theta < 2\pi$ $radians$ $causing$ the $constant$ $22/7.$ It also causes the velocity of the
space-time continuum to travel
at πc or $n\pi c$ $where$ n is any $real$ or $imaginary$ $number.$

The Schrodinger Equation describes the wave function of a point
particle
$\psi(r,t)$ in a $spacial$ $state$ of $three$ $coordinates$ at $time$ $t.$ The $probability$ $density$ is $d)(r,t)or d3r$

At time t for point r.dP(r,t)=C|$\psi(r,t)$|2d3r $where \int$ $dP(r,t) = 1$ $and\psi(r,t) =$
$\int$ $|\psi(r,t)^{2d^{3r}wit}h$ the $normalization$ $constant$ C $where$ $\frac{1}{c} =$
$\int$ $|(\psi(r,t)|$^2d^3r has c =$
1 as a $normalized$ $wave$ $function$ and $\psi(r,t)$ is $continuous$ $everywhere.$ So the $Schrodinger$

(13)NOTHING DOESN'T EXIST AND NEVER DID AND NEVER WILL

One of the biggest arguments in modern physics today is what happened before

"The Big Bang". In "The Big Bang ex nihilo" a quantum bubble appeared out of

nothing then exploded 10^{-43} seconds later or Planck Time to create our universe.

Nothing by definition has no properties so that a place holder like true 0(zero)is still

not nothing as it has the property of being a place holder. Spacelessness is

impossible(that which is beyond the comprehension of mankind) and the lack of

potential energy of space and the kinetic energy of the Cosmologic Constant

wouldn't occur. Even if mass created space around it that mass has to come from or

emanate from energy which is a property of the massive Higgs Field and Higgs

Boson below the speed of light and the tachyon above the speed of light. "The First

Event" was a time oscillation paradox between the Higgs Field below the speed of

light and a rogue group or grouping of tachyons at the speed of light with times'

arrow moving in both directions simultaneously causing a spin(forming magnetic

moments and the monopole 1-Brane and 2-branes respectfully. The spin became a centrifuge effect forming the spiral space-time continuum with the string

dimensions at the center of the spiral and the macroscopic dimensions at the

periphery. In addition space has potential energy from which the kinetic energy of

the Cosmologic Constant was formed that kept the D-0-branes from touching each

other becoming the D-1 and D-2 branes followed with the time oscillation by having

dilated time(very slow) from beoming the 4th dimension or the D-5 brane. There

were no boundaries at the extremes of the D-0 branes so length and width were

meaningless and depth was formed when the D-0 branes touched after being

separated by the anti-gravity of the Cosmologic Constant. As there were an infinite

number of D-0 branes(space-time is infinite) the cosmologic constant gets

multiplied by infinity forming Dark Energy which will eventually blast apart all

matter into fermions and possibly cause a rift in the gluons holding the fabric of

space-time together causing a "Big Crunch" for our universe or a huge spinning

effect in the fluid space-time continuum which encompasses all the universes in the

multi-verse. Nothing is the absence of energy (including potential energy and

kinetic energy noted as Gibbs Free Energy) as well as the absence of matter and space-time. Without space-time one cannot have energy or matter therefore one

cannot have a Higgs Field or tachyons. Also if time didn't exist (the sequencing of

events) one could not have a first event without a catalyst and no catalyst is

incorporated into the "null set" so spacelessness would be for an infinite period of

time which means we wouldn't exist and due to the weak Anthropic Principle we

exist so the sequence of events must occur for the existence of intelligent life and

spacelessness in non-existing time would preclude that possibility. There were

Quantum Fields with perturbations caused by energy and the formation of matter is

space-time. The energy is referred to as the zero-point energy level which is

discussed in greater length in my book" Mega-physics III; Nothing Doesn't Exist" and

"The Nth Power". The zero point energy encompasses the Cosmologic Constant and

the potential energy of free space inclusive of Gibbs 'Free   Energy which is an

integral formation of entropy and incorporates the kinetic energy of the Cosmologic

Constant as well as the existence of leptons(massless components of space-time

which form the orbifold in string theory) and gluons which encompass the Strong

Florce as a Superstrong Force holding space together intact unless or until Dark

Energy reaches a critical point where it fractures up and down quarks and then gluons to form a huge universe sized whirlpool in fluid space-time analogouds to a "Big Crunch". The point of creation was when a rogue tachyon or tachyons dropped to the speed of light forming bosons including the spin 2 vector boson and causing the Time Oscillation Paradox of the" First Event". As we exist nothing doesn't exist as it would still be nothing and a catalyst(time oscillation paradox) would be necessary to trigger "The First Event"

The peak wave length of Hawking Radiation is almost 16 times the Schwarz child

Radius of a Black hole. The equation

is $\lambda\ max = \left(\frac{8\pi^2}{4.9651}\right)\ r(s) = 15.902\ r(s)\ where\ r(s)\ is\ the\ Schwarzchild\ Radius$ of a

black hole event horizon and the wavelength of Hawking Radiation relates to flat

Minkowski Space-time as ds^2=dx^2+dy^2=dz^2-c^2dt^2 from the line element.

With regard to Spooky Action at a Distance Hawking Radiation is electromagnetic

radiation perturbing space-time in a spiral configuration so the spiral operator

$2n+1^\wedge\pi\omega\ i \rightarrow j\ (for\ expanding\ space - time\ from\ a\ point\ to\ flat\ space -$

$time\ for\ Hawking\ Radiation\ to\ perturb\ space -$

$time\ toward\ the\ adjacent\ black\ hole\ event\ horizons\ where\ space -$

$time\ constricts\ by\ the\ spiral\ operator\ to\ \frac{1}{2^{n\pi\omega\ j}} \rightarrow$

$i\ and\ the\ angular\ momentum\ increases\ geometrically\ as\ the\ event\ horizon\ of\ all\ other$

black holes are approached. The energy of Hawking Radiation is

$\rho(H)\ in\ the\ Equation\ of\ Everything\ is\ i\hbar\rho(H)\ so\ that\ 2\pi(R\ abc + or -$

$$\frac{1}{2} R \; g \frac{ab}{i} h\rho(H) \, and \; \Sigma\rho(\text{H}) =$$

$\lambda\,\text{max}$ *or the peak wavelength which is the reciprocal of the frequency υ which relates to the er*

energy of Hawking Radiation. As photon pairing is electromagnetic radiation and is

subject to Spooky Action at a Distance photons also follow the paths of space-time

curvature in the same way as other types of electromagnetic radiation as all are

comprised of photons including Hawking Radiation.

Utilizing the formula for space-time as -1/2e-I n cot(theta) where theta is the

trajectory of Hawking Radiation the trajectory is 90 degrees or

$\frac{\pi}{2}$ *radians which is* $\cos\dfrac{\frac{\pi}{2}}{\frac{\sin\pi}{2}}$ *which is* $\frac{0}{1}$ *or* 0 *. This is consistant with the constriction of space* $-$

time towards zero at the event horizon of any active black hole. Thus $=$

$\frac{1}{2} e^{-in(0)w}$ *heren is the two dimensional state relating to the flat matter of Hawking Radiation*

e^0=1 and 1/2e^0 =1/2 while -1/2e^0=-1/2 so the midpoint is the inception of a

self contained LOOP OF HAWKING RADIATION IN CONSTRICTED TO DILATED

SPACE-TIME FROM THE EVENT HORIZON OF ONE BLACK HOLE TO THE EVENT

HORIZON OF ANOTHER BLACK HOLE.THIS IS TOTALLY CONSISTANT WITH TOTAL INFORMATION EXCHANGE OF HAWKING RADIATION BETWEEN EACH AND EVERY BLACK HOLE WHERE TIME ISN'T IMPORTANT AS HAWKING RADIATION GOES FROM CONSTRICTED SPACE-TIME OF ONE BLACK HOLE WITH DILATED TIME TO CONSTRICTED SPACE-TIME WITH DILATED TIME OF ANOTHER BLACKHOLE THEREFORE THE INFORMATION DISBURSED INTO SPACE-TIME AS EITHER A SUSPENSION(where the 2 dimensional lattice formula can be applied)or if the information is dissolved in space-time and may travel in a formulation similar two the Robinson Congruence (which is a superfast pathway for photons which comprise electromagnetic radiation).These phenomena also explain photon pairing as a form of Spooky Action at a Distance .The phenomena involving spooky action with electrons was described in this author's second book "The Equation of Everything". 12.3 AN M THEORY APPROACH TO HAWKING RADIATION

Assuming that Hawking Radiation is comprised of electromagnetic radiation or photons which are flat or two dimensional matter this would be associated with the 3-Brane associated with two dimensional space ,the 4- Brane associated with three dimensions of space where the third dimension of photons would be string sized or

10^-33 cm. The gathering or distribution of charge in the 2-Branes   are generally

associated with electrons which also has electromagnetic radiation as magnetism is

formed from electromagnetic radiation as electrons are transferred as well as

positrons(anti-electron)and the fact that photons have a miniscule mass would

cause them to smear along multiple membranes or Branes as the velocity of

electromagnetic radiation would only vary at just above absolute zero when

photons vibrate in a lattice of Boso-Einsteinian condensate. Space itself would be

comprised of leptons and gluons and photons would pass through the fabric of

leptons as long as the gluons remained intact holding the leptons together and

would therefore form the 4-Brane,5-Brane,....N-Brane all the way to the super-small

Super-brane There were 252 separate states of matter in the average active black

hole according to Stephen Hawking and black hole entropy

$S=2$

$$\pi\sqrt{NQ1Q5} \quad where\ N = number\ of\ states\ Q1\ and\ Q5\ are\ the\ differential\ charges$$

.Spooky Action between black holes would have space altered such that the 3 Brane

would compress toward the superbrane constricting the volume of photons which

would increase the volume geometrically BY THE MUZZLE EFFECT AS WITH A

WATER BLAST IN A MUZZLE.CONSIDER ELECTROMAGNETIC RADIATION AS A

PERFECT GAS WITH THE PERFECT FLUID OF SPACE-TIME;CONSTRICTING THE

PERFECT GAS ENOUGH WILL INCREASE THE VELOCITY TO EXCESS OF 3×10^8

meters.sec as the velocity of space-time would also increase geometrically with

constriction nd the photon stream of Hawking Radiation would constrict with the

constricting space-time.Therefore photons would constrict to sizes which are

virtually immeasurably small and immeasurably fast in the massive Super-brane

explaining Spooky Action at a Distance for Hawking Radiation using M Theory.

FOOTNOTES;
1.Kaku ,Michio. Introduction to Superstrings and M Theory p.61-62 and section 1.8
Harmonic Oscillator
2.Peebles.Principles of Physical Cosmology p.500-503
3.Wikipedia.Hawking Radiation.

CHAPTER 13:

WHAT IS MORE LIKELY THAT THE BIG BANG WAS PRECEDED BY A BIG CRUNCH

AND WAS NOT THE SECOND EVENT BUT HAPPENED AEONS LATER OR THAT THE

BIG BANG WAS THE SECOND EVENT FOLLOWING THE FORMATION OF

STRINGS,STRING DIMENSIONS AND MATTER?

Utilizing Occam's Razor the first explanation appears on the surface to be more

plausible as it is logical and simple ,however ,is it? The second explanation would

place the site of the Big Bang at the center of the vortex formed by the Time

Oscillation Paradox acting on the infinite parallel planes of space. In order for the

second explanation to be more correct would have to illustrate a recoil effect or

spring effect in the opposite direction from the Big Bang which initially went

outward in all directions from the center of the spiral vortex of space-time which

had been previously formed. This spring effect would have had a deceleration early

on after The Big Bang after which time the acceleration outward in all directions

would have been a pull from space-time acting on space-time from our universe as a

current along with the push of dark energy which is a relatively weak force of anti-

gravity. In other words there was a push outward from Dark Energy from the

mutual repulsion of anti-particles in the quantum bubble from the center of the

vortex but this push was almost IMMEDIATELY overtaken by the pull of space-time back toward the center of the vortex causing the accelerated expansion toward the center of the vortex .Note immediately could be anything from Planck Time or 10^-43sec to several hundred thousand years. The answer is in the data .If the WOMP or any studies of gravity waves or measurements of space-time curvature would show a change or reversal in the acceleration early on followed by a substantial increase in the acceleration .As a consequence ,with the knowledge our technology has the conclusion must be soft or tenuous at best albeit possible. Theo other alternative which is simpler and easier to explain is a Big Crunch of another universe with space-time constricting toward zero and time dilating toward infinity (infinitely slow relative to the observer) which would mean the progressive dilation of time would slow either slow it toward zero from positive time moving progressively slower until time 0 is approached as the quantum bubble is formed in constricted space-time or time would move backwards from positive toward zero in a negative direction as in the case of a massive IMPLOSION. In this case The Big Bang was certainly not the second event but occurred much much later. Of course the collision of membranes could also have caused it but of course this too would mean it isn't the second event but would have happened much much later. So what is correct?

Are all things equal? If science can answer that question we will be considerably

closer to the OMEGA POINT. To learn all things that are learnable would answer the

question "When are all things equal? ".Until then the first statement toward wisdom

is "I don't know".

13.2 ANTIPARTICLE ANTIPARTICLE REPULSION AS CAUSE OF THE BIG BANG

IN A SYMMETRICAL 360 DEGREE ORB BLAST AT THE BIG BANG R^ijkl=antigravity
effect on particles and Rji^kl is the antigravity effect on antiparticles.$\Lambda ij = 1/Rij^2$
$=1/8\pi G$ where $R\,ji =$
$antigravity\,from\,antiparticles\,\dfrac{1}{Rij^2is}\,spacetime\,from\,the\,Cosmologic\,Constant.\,The\,vector\,prod$

$Rij\text{^}2 \otimes R\,ji\,reveals\,R^{2ijRji}/|Rij^2||Rji =$
$cos\theta\,which\,revelas\,that\,e\,ji\,\dfrac{R^{2ijR}ji}{||R^{2ij}||}\,||Rji||\,where\,0 < \theta <$
$\pi\,radians.\,In\,a\,symmetrical\,orb\,blast\,the\,re\,are\,an\,infinite\,number\,of\,slices\,with\,a\,total\,trajec$

tory of
$\pi\,radians.\,The\,cosine\,of\,\pi\,radians\,is\,-1.\,so\,R\,ji =$
$\dfrac{1}{8\pi G \otimes g}\,ij\,The\,vector\,product\,of\,Rij(R\,ji) = e(ij,ji)cos\theta.\,Rij^{2(R\,ji)} =$
$1.\,For\,\pi\,radians\,R\,ji = -e\,ji.\,As\,g\,ij\,is\,the\,metric\,of\,a\,particle\,and\,-eji-$
$is\,sspacetime\,curvature\,or\,gravity\,of\,R\,ji.\,g\,ji\,is\,the\,metric\,of\,the\,antiparticle.\,R\,ijkl\,is\,spacet$

Time for antimatter in four dimensions and kl is positive space-time. With R
jikl=covariant tensor and Rji^kl being the contra-variant tensor R ji=-e ji and R jikl-
Rji^kl=-eji=g ji/8(pi)G where R jikl-R ji^kl is the space-time curvature from anti-
matter,R jikl-R ji^kl=-
$8\pi G\,eji = g\,ji$ where $g\,ji$ is the Riemann metric for an antiparticle and $-$
$8\pi G$ is the ANTIGRAVITYFOR ANTIPARTICLES IN ANTIMATTER ILLUSTRATING A MUTUALLY F

REPULSIVE FORCE.R ijkl is space-time curvature for gravity and R jikl is space=time

curvature or reciprocal curvature for antigravity and of course g ij is the metric for

particles with positive gravity and g ji is the metric for the anti-particle with anti-

gravity. Finally R ^ijkl is the antigravity effect on particles and R ji^kl is the

antigravity effect on antiparticles which forms the -

8

πG which is the multiplier of e ji to equal the metric g ji which is the metric of the antiparticle

Clearly the negative sign illustrates anti-gravity for anti-particles.

13.3:EVERY ACTION MUST HAVE AN EQUAL BUT OPPOSITE REACTION
 According to Sir Isaac Newton (NEWTON'S THIRD LAW) every action must have

an equal but opposite reaction. The Big Bang ex nihilo breaks that law of motion;

however with the spring effect of the quantum bubble from emerging   strings and

string dimensions from the center of the vortex of spiral space-time IS THE ACTION

after the time oscillation paradox spuming out in all directions to the point of

deceleration from the center of the vortex. This then reverses toward the center of

the vortex again with the accelerated expansion of our universe. Again this is due to the enormous pull of space-time from the vortex on space-time and push of dark

energy within our universe and is the REACTION.. The equal but opposite reaction

to the expansion of our galaxies from the quantum bubble and Big Bang was the

push of anti-particle anti-particle repulsion within the quantum bubble from the

center of the vortex to the POINT OF REVERSAL OF THIS UNIVERSE BACK

TOWARDS THE CENTER OF THE VORTEX OF SPACE-TIME. Of course a Big Crunch

of another universe with the same space-time current as our universe would also

obey Newton's Second Law. Therefore the event of "creation" in all actuality

occurred when the rogue tachyon (s) dropped to the speed of light initiating the

time oscillation paradox on the infinite parallel planes or the D-0-Branes. Again it is

unclear as to whether only one universe or the multiverse formed from the advent

of string dimensions and strings as well as the spin 2 vector boson from the rogue

tachyon(s)initiating gravity and converting almost 100% potential energy of the

components of space into 10^77 joules of kinetic energy after the paradox acted

upon these infinite parallel planes spinning them into a centrifuge effect forming the vortex of space-time with matter gravitating towards the center as an origination

point. THIS IS CLEARLY NOT OUT OF NOTHING AS NOTHING DOESN'T EXIST.

13.4 WAS THE ROGUE TACHYON COMPONET OF THE HIGGS FIELD (PRIOR TO

THE FORMATION OF THE MASSIVE HIGGS BOSON) A PURPOSEFUL ACTION?

 To mathematically determine if the rogue tachyon (s) was impelled by a conscious

force(determinism vs. free will) is bringing together science and philosophy

merging into religion and while K-suryon waves were determined to

mathematically exist; it still hasn't been conclusively proven that consciousness is

energy. Until that question is answered it will be unclear if creation was purposeful

or random(which according to Quantum Mechanics would have a 100% probability

with infinitely or nearly infinitely dilated time) for that event to occur. Chaos Theory

would indicate randomness which follows the laws of entropy which would indicate

that the most ordered state would be prior to the first event. Again if the vortex of

space-time relates to the vortex of a black hole then the entropy would approach

zero in the center of the vortex but not be zero. Stephen Hawking stated that black

hole entropy approaches 0.29 and it is possible that the entropy of the quantum bubble would also approach that value in the center of the vortex of space-time with

252 disparate states of matter and space-time constricting towards zero without

reaching it at the center(BEING THE LOCATION OF THE BIG BANG).

13.5 THE DISTRIBUTION OF SPACE-TIME AFTER THE FIRST EVENT

Space-time acts as a perfect fluid and the spiral configuration of the space-time

continuum is like a giant whirlpool whose outer edges stretch out toward infinity in

all directions from the center of the vortex. As the distance from the center of the

vortex increases in all directions over 360 degrees the amount of curvature

reduces until at the extremes (which are never met) space-time approaches

absolute flatness which is like a stagnant pond. The quantum ground state as

described by the spiral fractal formula is the ZERO DIMENSIONAL STATE and 100%

potential energy-

$\epsilon(negligible\ kinetic\ energy)is \sim zero\ energy\ but\ is\ actually\ \varepsilon\ kineitic\ energy\ and\ 100\% -$

$\epsilon\ potential\ energy\ from\ the\ components\ of\ space.\ \int_0^\pi \frac{du}{u} = \ln u \to 0.$VACUUM

ENERGY IS THE SUM TOTAL OF THE POTENTIAL ENERGY OF SPACE WHICH IS

COMPRISED OF LEPTONS AND BEING HELD TOGETHER BY MASSLESS GLUONS.IT CAN BE CONSIDERED AS 10^77 joules incurred by the Time Oscillation Paradox of the first event. Of course as mentioned in this author's previous book "What is the Dimension of Time?" time is the sequencing of events and without time all events would occur simultaneously in the same space which absolutely disallows space-less-ness as space-less-ness cohabiting space containing everything with sequencing would implode everything into space-less-ness which would require time as an implosion is an event as would be the formation of space from space-less-ness therefore they would have been nothing without time which can only be nothing and could have only been nothing WHICH IS NON-EXISTENCE therefore once again nothing doesn't exist and time does.

The classical definition of potential energy is P.E.=m g h where m=mass g=gravity and h=height .In the infinite planes hypothesis the height of the infinite planes is infinity. The mass of leptons is approaching mass -less-ness but still has a miniscule negligible mass and therefore negligible gravity which bleeds through from tachyons. Gravity is the curvature of space-time caused by mass   ,but the curvature of space is from time constricting or dilated space. If time is infinitely or nearly

infinitely dilated the constrictions of space by time approaches ZERO so the pre-

first event space is flat and totally parallel .As a consequence the negligible mass

causing gravity cannot curve space with time completely dilated as it is time that is

actually constricting space. As a result

$$mass = \epsilon \quad gravity = \epsilon \ and \ space - time \ curvature =$$

$$0 \ as \ time \ is \ dilated. So \ the \ potential \ energy \ is \ (\epsilon)(\epsilon)(\infty) =$$

$$\infty \ and \ there \ is \ therefore \ \infty(infinite) potential \ energy \ which \ released \ 10^{77 \, joules} of \ kinetic \ energy$$

from the first event. Of course the Newtonian Definitlon of Potential Energy as

weight times height where weight is mass times gravity is only an approximation as

it may not consider Relativistic effects on potential energy.

CHAPTER XIV(14) HOW DOES THE HIGGS FIELD PLAY IN WITH THE FIRST EVENT

The Higgs Field is comprised of the massive Higgs Boson and TACHYONS WHERE

TACHYONS ACT AS A CENTRAL NERVOUS SYSTEM ND THE HIGGS FIELD ACT AS

THE BODY WITH DARK MATTER ACTING AS A WEB RELATING TL THE

PERIPHERAL NERVOUS SYSTEM.

Creation has a grouping of tachyons drop to the speed of light from greater than

light speed forming the Time Oscillation Paradox involved with the creation of the

spiral space-time continuum. The Higgs Field is postulated to transform energy into

mass which cannot be performed otherwise in this universe. The Higgs Boson was

isolated in Cern Switzerland and the potential energy of free space or leptons and

gluons can be transformed into kinetic energy equaling 110^{77} joules. Gravity is the

curvature of space-time caused by mass and is an effect as is anti-gravity.

CHAPTER XV (15)

WHAT ARE" BLACK "HOLES AND HOW DO THEY RELATE TO" WHITE" HOLES

Black holes are area of un-delineated mass which lack any reflected light at all. Although referred to as "black" there is an absence of any reflected light in that area of delineation. All light is absorbed due to the extreme curvature of space-time as the event horizon is approached time is dilated by the extreme mass of the black hole(mass always dilates time or slows it down)and the dilated time constricts space as the event horizon is approached. Gravity is the effect of curvature of space-time caused by mass and the extreme mass curves space-time to almost a point of infinite curvature.

White holes reflect virtually all the light in the delineated area and are more scarce than black holes in our universe. This indicates an area of extreme anti-gravity or the curving out or flatten of space-time caused by the pushing out or lack of cohesion of mass in this case anti-matter which lacks cohesion due to the mutual repulsive forces of each anti-fermion (although in Cern Switzerland anti-protons and poistrons(anti-electrons were produced with enough positive gravitational effect(not force) to reverse the reciprocal curvature of space-time caused by the

mass of the anti-matter. In other words the mass of positive matter can curve space-

time inwardly enough to cancel out the outward curvature of space-time caused by

the mass of the anti-matter increasing it's cohesion enough (although temporarily)

to exact measurements. Anti-matter conduits may actually accumulate in "white

holes" and the anti-gravitional effect may bend light to almost 100 per cent

reflection. It is also possible that white holes may be an "exit point" to black holes

with the center of the vortex of any black hole being the center of two mirror images

with opposite effects exiting in other universes.

HOW CAN A C.P.T. VARIANCE CAUSE A REVERSAL IN TIMES' ARROW?

According to Schrodinger's Equation and Quantum Field Theory the partial

derivative of the wave function of any point particle with respect to time is equal to

the Hamiltonian Operator operating on potential energy plus potential energy of the

wave finction with respect to time. This assumes time to be C.P.T. variant with

respect to the C.P.T. theorem(Charge ,Parity, Time). If times' arrow is in the opposite

direction the C.P.T. theorem has an invariance however Potential Energy relates to

the wave function by way of the Hamiltonian Operator

$\frac{\hbar^2}{2m}$ $\nabla^n where\ \nabla\ is\ the\ LaPlacean\ Operator\ based\ on\ that\ degree\ of\ parial$

differential equations relating to the number of dimensions of the manifold in

question. In the case of Planck's Constant divided by 2 pi squared divided by twice

the mass in question times the La Placean Operator operating on Potential Energy of

the system one assumes time's arrow to be in one direction. When the wave function collapses the equation of the partial derivative of the wave function with respect to time approaches zero as in the case prior to the First Event. If the energy of a photon becomes negative but the frequency remains positive Planck's Constant becomes negative and it's square remains positive. The question then arises when is potential energy negative? When all energy is kinetic or heat and there is virtually no mass and space is constricted potential energy of space has been converted in kinetic energy and heat and space-time emanates as a severe flattening effect or reciprocal curvature as if in a White Hole is the White Hole contained anti-matter where all fermions repel each other. There is a postulate stating that there exists a paralle universe to ours where times' arrow points in the opposite direction to ours. If the "white hole" in our universe converts to a "black hole" in a parallel universe times' arrow may be in the opposite direction to ours. It is also possible that that universe has cohesive anti-protons ,positrons and anti-neutrinos as well as anti-neutrons however bridging the gap between universes can kill a time traveler whose inner clock is geared to have time move in one direction while the environment moves in the opposite direction. This was discussed in my book The

Nth Power. If the mass in the Hamiltonian Operator reflects as negative and the n

dimensions associated with the La Placean Operator are an odd number(our

universe has it as even with 26 dimensions compactified (curled up) to 10 which

becomes 4 macroscopic and 6 string dimensions) ;the Charge Parity Time Theorem

is no longer invariance for –time to exist with time's arrow pointing in the opposite

direction.

CHAPTER EIGHTEEN

BLACK HOLES AND RELATIONSHIP WITH DIFFERENT STATES OF MATTER   AND
BLACK HOLE ENTROPY

Matter and energy are entangled under some specific circumstances .Are strings

flat matter which is 2 or possibly 1 dimensional or are they 0 dimensional as energy

only with motions tension and vibrations as only frequencies.Is there a Law of

Conservation of Matter as matter and energy are inter-changable and entangled

under the extreme pressures and temperatures in a black hole? According the

Schwarzchild   Space-time there is a constriction of space-time with time dilating to

almost infinity or in essence stopping at the Event Horizon of a Black Hole. Black

holes emanate from collapsing matter in galaxies, neutron stars and possibly

universes. As a consequence there should be a black hole at the 0,0 point which is

the point of the "Big Bang" although in an isotropic universe scientists may not be

able to locate it for a considerable period of time The initial rotational vectors in the

expanding universe where rotation is progressively decreasing as space-time

continues to push outward should be slightly measurable as the proximity to this 0,0

78

black hole is approached although iso-tropism generally precludes a center of gravity. Still it makes sense that the 0,0 point would be a center of rotation for the rotating and expanding universe(this universe in the multiverse)It is now postulated that all galaxies have a central black hole including the Milky Way and these galaxies rotate along the central axis of these black hole albeit at an extremely slow rate.If negative mass exists in a black hole it was mathematically determined in chapter of Schwarzchild Space-time that superimposition of region 3 on region 2 and regions 1 and 4 can cause the information to garner at the Event Horizon like a phonograph record or video tape .This relates to the idea of The Holographic Universe and translates everything into two dimensions .Regions 1 and 3 are left and regions 2 and 4 are cancelled. Region 3 has negative mass and superimposes on region Region 3 is reflected back on region2 with region 3 having a negative mass and region 2 having a positive mass and there canceling as per the math in chapter 5,leaving regions 1 and 4.This may be a solution to the Hawking Paradox which says that as a black hole evaporates information in the black hole is lost.

Regarding the 252 different states of matter within a black hole(Hawking) these

would have to include all or most state under extreme pressure with significantly

constricted space and extremely high density .If electromagnetic radiation including

photons from light are absorbed into a black hole(making it invisible)would that

radiation show matter or matter like characteristics such as the states of liquid

radiation suspended in a condensate as in the experiment by Dr. Len Hau mentioned

in the previous chapter .If radiation could occupy the different states of matter and if

matter and radiation can be or are entangled then radiation can occupy a liquid and

possibly a Boso -Einsteinian Condensate state at near 0 degrees kelvin. If this is true

perhaps radiation can in the future be considered a "perfect gas" but again for that

one would have to demonstrate mass in a photon although of course electrons and

photons have mass as do anti-protons and positrons although the mass in the latter

two may be considered "strange mass" which might be an oscillating hybrid

between negative and positive mass although recorded as positive mass which may

or may not have anti-gravitational effects rather than the effect of pure gravity.

80

In terms of The Equation of Everything" as space-time curves in manner reciprocal

to ordinary space or curves inward when in the expanding universe space-time

curves

outward $\mathbb{R}n = R\,abc - \frac{1}{2}R\,g\,ab \div R\,ab$ or $\mathbb{R}n = \prod n = 1\ to\infty\ \frac{1}{2^n\pi}\,g\,ab\,R\,abc -$

$\frac{1}{2}R\,g\,ab \div$

$\rho\,ab$ where the infinite product of $\frac{1}{2^n\pi}$ times the metric $g\,ab$ times Mintkowski Space $-$

time divided by the energy density of matter for the metric $g\,ab$ reveals for a large mass

In constricting space reaches a point where $-1/2R\,g\,a\,b = R\,a\,b$ where the space-time

curvature variant of the mass $R\,a\,b =$ inertial mass of the black hole such that

$\mathbb{R}n = R\,abc \otimes 1$ where $R\,abc$ constricts with near infinte curvature from $R\,g\,ab$

and $R\,ab$ is a huge number. The expression $\prod \frac{1}{2^n\pi} \to \frac{1}{n\pi}\,g\,ab = \frac{1}{\infty} = 0\ space -$

time with near infinite curvature in a black hole with constricted space. Here $\mathbb{R}n \to$

$0\ as\ n$ approaches a large number. The expression $\rho\,ab$ relates to $R\,ab$ as the enrgy equivalent

of the inertial mass R a b and incorporates the 1/c^2 or 1/p^2c2+m^2^c^4=energy
and p=momentum incorporates into

ρ ab the $energy$ $density$ $with$ $regard$ to the $metric$ g ab.Of course the

1/2

π $relates$ to the $spherical$ $nature$ and $circumference2\pi\rho$ at the $event$ $horizon$ of a $black$ $hole$ wl
space-time is constricted down toward 0 with infinite curvature and as the areas
where the energy density of

matter

ρ and $inertial$ $mass$ $constrict$ $more$ and $more$ it $goes$ $from\frac{1}{2\pi}$ $to\frac{1}{4\pi}$ $to\frac{1}{8\pi}$ $times$ the $metric$ g ab an

stress energy T

$ab\rightarrow 1.$ as $inertia$ $\leftarrow$ $gravity.As$ ρ ab $\rightarrow$

$large$ $number$ the $entire$ $expression$ for $\mathbb{R}n$ $\rightarrow$

0 but $where$ $does$ the $energy$ go $from$ the $event$ $horizon$ of a $black$ $hole.$ $Answer$ $quasars$ $with$
Hawking Radiation being spumed out like a jet engine leaving the black hole cold as
a C02 cartridge would be cold after the contents of the cartridge were suddenly
forced out

.Therefore

ρ ab $would$ $manifest$ the $energy$ for the $inertial$ $mass$ R ab $like$ a $CO2$ $cartidge$ $being$ $discharge$

over a protracted time period.
The expression

S=2

$\pi(NQ1Q5)^{1/2}describes$ $black$ $hole$ $entropy$ in $terms$ of $it's$ 252 $different$ $disparate$ $states$
.This would be similar in some ways to $2\pi\rho = circumference(space - time) \rightarrow$

0 so $\rho \rightarrow 0$ $deep$ $within$ a $black$ $hole$ $ and the diminishing sphere of space-time to a

point is described by$\mathbb{R}n = \frac{1}{2^{n\pi}as}n \rightarrow \infty$ $gab\otimes Rabc - 0.5Rg$ $ab\otimes\rho$ $ab^\wedge - 1$ where

space-time=$2\pi\rho$.
Note that a circle reducing to a point is a cone or asymptotically a spiral. Therefore
the operator $\Pi n = 1$ $to\infty$ $½^\wedge n\pi$ would be the spiral operator on space-time
reducing it from a sphere to a point in a cone shape The Spiral operator would be

$\Pi\frac{1}{2^n}$ $\pi^{-1}where$ the $infinite$ $product$ Π $ is$ $from$ n to ∞ eigen-states.

The Bekenstein Hawking Equation for black hole entropy
S=A/4Lp^2c^3A/4G$\hbar$ $where$ $\hbar =$

$Planck's Constant6.63x10^{-34joule} - sec$ $ $ $ or$ $meters^{2kg}/sec$ $G =$
$gravitational$ $constant6.67x10^{-11}newton$ $meters/sec$ $^2 Lp = Planck$ $length =$
$10^\wedge - 33cm$ A=cross sectional area or kA/Lp^2 where k=Boltzman constant and

Pl=$\hbar\frac{G}{c^3} =$

$10^{-33cm}again.$ $Black$ $hole$ $entropy(S)$ $based$ on $this$ $is\frac{1}{4}$ or $also$ $Steven$ $Hawking$ $postulated$

0.29.Black hole entropy is directly proportional to the area of the event horizon by
the Boltzman Constant k which was explained earlier with the Boltzman Equation
for different states of matter.This is the maximum entropy obtained by what's called

82

the Berkenstein Bound and relates to the Holographic Universe and principle of a
two dimensional fingerprint of information in the black hole at the event horizon.
Supersymmetry was applied to black holes using D-branes and string theory duality
with regard to the SO(32)string theory and the compactification of closed string
theory IIa to a circle with regard to M theory.
The zeroth law states the surface gravity of the event horizon of a stationary black

hole doesn't vary. The first law states that the the change in energy dE=k/8πdA +

$\Omega dJ + \Phi dQ$ where $\kappa = surface\ gravity\ A = area\ of\ event\ horizon \Omega =$

$angular\ velocity\ J = angular\ momentum \Phi =$

$electrostatic\ potential\ of\ the\ charge\ Q$ The second law is that the horizon area is

a non- decreasing function with regard to time or dA/dt> or =0.Whenit was

discovered that Hawking Radiation was emitted by black holes and the area and

mass(therefore gravity) decreased over time this was became a" weak law". The

third law of black holes is that

k

$or\ \kappa\ for\ surface\ gravity\ cannot\ be\ 0$ $The\ zeroth\ law\ states\ that\ surface\ gravity\ is\ similar$

to thermal equilibrium in thermodynamic systems in that a temperature doesn't
vary neither does surface gravity $\kappa in\ a\ black\ hole$.Wikipedia;Bekenstein Entropy

In the equation S(BH)=2$\pi\sqrt{NQ1Q5}$ N is the number of states as mentioned before
and Q1 relates to the one-brane which relates to effecting or carrying the electrical
charge related to the monopole or the electron emanating from higher states of

matter in terms of n-branes.Q5 relates to the charge relating to space-time as the electron has a cloud with a probability density that goes out to ∞ *without reaching it but bounded by the event horizon of a a black hole.*

Based on this $S=2\pi\sqrt{(252)}$ *charge of an electron gas(space $-$ time)*

Charge of an electron is 1.6x10^-19 coulombs and the 5brane relates to near infinite curvature of a contracted area of space-time as told by

$$\frac{\Pi 1}{2n\pi}\ g\ ab\ R\ abc - \frac{1}{2}r\ g\ ab \otimes \rho\ ab^{-1} where\ n =$$

number of eigenstates in n dimensional space. for the infinite product from n to ∞ *giving spa*

½^nπ *as* $\frac{1}{2^{0}\pi}$ *as space $-$ time approaches the D $-$ 0 state or D $-$ 0 $-$*

branes making $\frac{1}{2\pi} = \frac{1}{2\left(\frac{22}{7}\right)} = \frac{7}{22(2)} = \frac{14}{22}$ *so 44/*

$$7\sqrt{252(1.6x10^{-19})\left(\tfrac{14}{22}\right)}=6.14[[(160.36)(1.6x10\text{-}19)]\text{^}1/2 \text{ which is } 6.141(6x10\text{-}$$

17=36.6x10^-17 as black hole entropy in the zero(0)dimensional state or the D-0-

brane which is very close to 0 entropy while Steven Hawking postulated entropy of

a black hole to be approximately 0.29. This is because while spacetime is constricted

at the event horizon it still exists so there is no 0 dimensional state within a black

hole which would make the 5-brane relate to a four or perhaps higher dimensional

state with the 252 different states of matter including energy-matter conversion or

entanglement due to the superhigh pressures and super cold temperatures. It is

likely that Boso- Einsteinian Condensate would exist with a matrix that traps

84

photons, electrons, positrons ,neutrinos ,anti-neutrinos ,bosons and at near the zero

dimensional state fermions. It is also possible that tachyons would be trapped in a

black hole which would reverse time's arrow and have a negative mass or strange

mass hybrid between negative and positive mass.It is these tachyons that would

cause the –mass in the superimposed region 3 onto region 2 that would solve the

Hawking Paradox. Whether radiation actually takes on a liquid state is agnostic but

not impossible under those conditions.

If the spiral operator operates on the function R a b c-1/2Rgab$\otimes$ $\rho ab^\wedge - 1$ and is inclusive of the metric g ab over the infinite product of eigenstates over n dimensional space it is asymptotic to Schwarzchild Space-time if the expression is reflected at the event horizon to express the increasing cone or spiral from approximately zero(0)dimensional space-time to n dimensional space-time where n goes from the D-0 eigenstate to D-n-eigenstate.This can be accomplished if $\prod from\ n = 11\ or\ n = 10\ decends\ to\ n =$
$0\ for\ the\ \ D -$
$brane\ which\ gives\ \ an\ opening\ spiral\ or\ cone\ past\ the\ event\ horizon\ as\ there\ is\ a\ descending\ co$

Cone at the event horizon with space-time spiraling in from ordinary space.The factor relating to the mass equivalent on both sides of the event horizon emanates from the function on which the operator is operating or "The Equation of Everything" so Schwarzchild
spacetime=$\prod n \frac{1}{2^{n\pi}} -$
$\frac{\prod n 1}{2^{n\pi} g} ab\ where\ the\ limits\ of\ the\ infinite\ product \prod\ are\ from\ n\ to \rightarrow$
$D\ 11?\ and\ from D11 \rightarrow n\ where\ n\ starts\ at\ D = 1\ not\ D =$
$0\ as\ the\ spiral\ is\ a\ modified\ cone\ acted\ upon\ by\ the\ metric\ g\ ab\ as\ per\ Rabc -$
$\frac{1}{2R} g\ ab \otimes (\rho\ ab)^\wedge - 1.$

CHAPTER NINTEEN

DARK ENERGY AND DARK MATTER

It was questionable about whether dark energy and dark matter were related until the missing mass calculation for the mass equivalent of dark energy and the mass of dark matter were found to be congruous. As so they share charcaterisitcs and can be considered in the same gauge symmetry group.This can be negociated with the mechanism of the "Big Bang" which is like lighting a match. From a chemical standpoint(as previously mentioned)the unlit phosphors would be the homogeneous antimatter-matter mix,the Big Bang would be the light being struck,and the charred residual would be the dark matter while the energy expended would be the Dark Energy coupled with the energy from the annihilation of matter and antimatter in the "Big Bang".

As mentioned previously the antiparticle-antiparticle repulsion o\in the quantum bubble under extreme pressure and temperature force a huge anti-gravitation force to push matter with space-time outward after the rotational component of "The Big Bang" had almost completely slowed toward 0. The anti-gravitational force pushing galaxies apart from each other is Dark Energy and the residual antimatter that was burned out is dark matter.As the anti-gravitational moment of the interaction was carried off by the Dark Energy ,Dark Matter has a gravitational effect instead of antigravity although it emanated from anti-matter.

As mentioned previously the particles of dark matter are on and about everything just as the BMR from the "Big Bang" is all pervasive ,but the particles of dark matter are so small yet homogeneous that they might be a multiple of Pl=

$$\hbar \frac{G}{c^3 or} 10^{-33}\, cm$$ *which would make the density of this superfine powder or dust have a very sli*

slight measurable density in the massive volume of space although the gravitational effects from the mass are considerable although measured indirectly. This too was already mentioned.

It has been postulated that Dark Matter is composed of baryonic particles and neutrinos possibly with anti-neutrinos and anti-Hadrons as a residual from anti-matter and that Dark Matter acts as a type of "cosmic glue" which has been present since the "Big Bang"

There is a question as to how much of the cosmologic constant

Λ *is related to Dark Energy, but the force to push the 750 billion galaxies apart from each oth* seems to be far greater than what was empirically measured for Dark Energy and greater thanΛ(*the cosmologic constant*)*which relates to* $8\pi G/c^4$.

With a stretch of the imagination and some creative math one can see that the 1st,3rd,5th and all odd dimensions have dark matter sequestered as it's gravitational effect can only be indirectly measured and mass in the even dimensions 2nd,4th,6th,8th etc. can have it's properties directly measured .This can evolve from the space-time formula of-1/2e^-i n cot

*θ where θ = π radians which is the trajectory of the "Big Bang".*Odd powers of i

86

give results containing I while even powers of I give real numbers such as i^2=-1 ,i^4=-1 etc while i^1=I and i^3=-I and i^5=-I such that if this was the scenario due to the inert nature of dark matter its mass may be postulated mathematically mass of dark matter=mass of ordinary matter/i $or\sqrt{-1}$.This would be likely if dark matter which is anti-matter based showed anti-gravitational rather than gravitational effect as Fg=Gm1m2/r^2 where 6.67x10-11n-m/sec^2 or 10-11=G and m1=mass/i m2=mass2/i such that Fg=Gm1m2/i^2r^2=-1(G m1m2/r2)which would indicate mutual repulsion of antiparticles and anti-gravity .While it may be true that the odd dimensions may have dark matter and possibly most of the missing mass equivalent of Dark Energy forming the difference between what is measured and mathematically calculated as a prediction on we will still have to show strange or oscillating mass in dark matter(as burned out antimatter).This oscillating property would be between + and – mass and while experiments with the Hadron Collider in Cern, Switzerland are looking for anti-gravity between anti-particles and particle-anti-particle interaction before they annihilate ,it has not clearly been demonstrated yet that anti-particles display pure anti-gravity although they do display some properties of anti-gravity.It also has not been demonstrated that anti-particles display anything except a positive mass.The question is this .How does experimentally show burned out antimatter in dark matter when it's effects can only be indirectly measured. Also is mass/i the same as strange or oscillating matter with a hybrid between +mass and –mass. If this property follows antimatter and antiparticles it may also follow dark matter with much huger antigravity with Dark Energy. Mathematically shadow odd dimensions cannot be ruled out but whether or not they contain the mass of dark matter or mass equivalent of dark energy is up to speculation.

CHAPTER 19

IS THE FABRIC OF SPACE COMPRISED OF STRINGS?

The fabric of space is comprised of virtually massless leptons and gluons holding the

fabric together. Space existed before the FIRST EVENT DURING THE TIME

OSCILLATION PARADOX WHWEW SPACE-TIME DEVELOPED A SPINS AND

CENTRIFUGE EFFECT FORMING A VORTEX IN WHICH THE MULTI-VERSE OR AT

LEAST OUR UNIVERSE WAS AT THE CENTER . STRINGS EXISTED AFTER THE FIRST

EVENT WHEN GRAVITY FORMED WITH THE SPIN 2 VECTOR BOSONS EMERGING

FROM THE FERMIONIC OR VACUUM STATE FROM WHICH THE FERMIONS OF

LEPTONS AND GLUONS FORMED BOSONS AND QUARKS) which entail the STRONG

FORCE)and are comprised of gluons which always existed. STRINGS HAVE A MASS

AND AS A RESULT ARE CONSIDERED MATTER OR ENERGY; NATURALLY THIS

ENERGY IS POTENTIAL AND KNIETIC ENERGY; yet the cosmologic constant which

repelled the infinite parallel planes preventing them from forming dimensions by

their touching always existed as the energy density of a vacuum is the cosmologic

constant. As a consequence the cosmologic constant always existed as did space. If the components of space are truly massless then it is not necessarily comprised of

strings and they formed after the first event, but if leptons and gluons have a

miniscule mass almost immeasurable(as in the case of photons) then they could

indeed be comprised of strings and strings could have always existed. If strings are

composed of pure energy then they existed before the first event but if they are

compromised of matter and energy or pure matter then they existed only after the

first event. As strings have a miniscule mass at least part of the string is composed

of matter but photons also have miniscule mass and photons are electromagnetic

radiation so energy must at least partly composed of matter. Even space with

potential energy have miniscule kinetic energy so there is an asymptotic value of

matter involved in leptons and gluons which comprise and hold together space. The

cosmologic constant reflecting antigravity always existed and was caused by the

miniscule mass of space but if gravity existed before the first event it is so small that

it approaches zero without reaching it, Based on this one must compare the

miniscule mass of each string with the miniscule mass of leptons ,gluons and

photons. If the mass of strings is greater than that of leptons or gluons and possibly photons then STRINGS OCCURRED AFTER THE FIRST EVENT ONLY,but if less THEN

STRINGS ALWAYS OCCURRED AND WHERE EXISTANT BEFORE THE FIRST EVENT.

CHAPTER TWENTY-NINE (29)

HOW THE FRIEDMANN EQUATIONS FIT INTO THE HUBBLE EXPANSION RATE

$$a{:}\ \frac{}{a} = -\frac{4}{3}\ \pi\ G\ \left(\rho + \frac{3p}{c2}\right) + \Lambda\ c2\frac{}{3}\ -\ \frac{8}{3}\ \pi\ G\ \rho\}\ R^{\wedge}2\ \ where\ by\ R\ ^{\wedge}2 = -kc^{\wedge}2$$

This particular equation is based on the area of a sphere
=4/3
$\pi\,r^3$ *assuming homogeneity   and a perfect sphere which doesn't follow inflation.*

T exp=1/H$\sqrt{3c2/8\pi G}$ where
$8\pi G$ is the gravitational coupling constant and H is the Hubble expansion coefficient which

Has recently been found inaccurate.. In all actuality
8
$\frac{\pi G}{c4}$ *is the coupling constant and* $\frac{\Lambda c2}{3}$ *increases exponentially with the dimension of time*

dilating space toward flatness.r3=-G$(\rho + \frac{3p}{c2} + \Lambda\ c2/3.$

Spacetime in "The Equation of Everything" is $\Gamma\ (R\ a\ b\ c\ d)1/\sqrt{N}$ $\sum_{\Lambda}^{\mu} N$ where

N=free energy states or eigen-states from the ground state or Cosmologic Constant

to the maximum excited state of Free Energy or mu as determined with the

Boltzmann Equation. Space is dilated by constricting time (faster time) and the

maximum Free Energy or mu is approached by Dark Energy as time dilates space

towards flatness. The rate of change of space is determined by the Friedmann

Equations based on the acceleration coeffiicents and the homogeneity of the area of

our universe as assumed to be a perfect sphere(which it isn not).)

CHAPTER 22

IS THIS UNIVERSE TWO DIMENSIONAL?

Flat matter consists of photons ,strings(open and closed)and anything else

perceived as a holographic projection. The 2 dimensional lattice equation of Michio

Kaku indicates a relationshipship between quantum mechanics and relativity with

regard to the critical exponent when everything comes apart. Could this relate to

"The Big Bang" or "Big Crunch'?

The curling up or compctification of string theory or M Theory is a circle not a

sphere and the dimensions which are greater than two in a holographic construct of

the world sheet are compactified dimensions and string sized forming membranes.

White noise utilizing CMB data contains trillions of bits of information which are

string sized 10^{-33} cm or smaller In this author's first book "Megaphysics,A New

Look at the Universe"(2003)it was postulated that a two dimensional universe

would exist as would a world sheet in string theory helping to unify quantum

mechanics and relativity. This would include space in dilated time comprised of almost massless photons of electromagnetic radiation which would travel at all

speeds except zero depending on the medium they pass through and because

influenced by gravity such as in black holes they do have miniscule mass as does

space. So in essence the vacuum or fermionic state of matter may be comprised of

photons with gluons holding space together ,while tachyons still exist above the

velocity of photons traveling backwards in time which would be infinitely slow if

time were totally dilated. The event horizon or any active black hole contains bits of

information which would or might appear to the observer as being two dimensional

as well as in dilated time and photons are postulated to have existed virtually

always. In addition strings whether open or closed are comprised of flat matter or

energy which is also flat matter and therefore two dimensiona lThe critical

exponent of the 2D lattice equation may relate to the Big Bang w ith reference to

criticality as might a Big Crunch.

'The Holographic Universe 'by Dr. Micheal Talbot indicates that two dimensional
projections would exist in terms of the World Sheet as purported by string theory as
strings are two dimensional. Paranormal phenomena like "ghost sitings"could be
two dimensional images on a background three dimensional plane as "ghosys" are
indicated as "white noise" or in the low energy fradio wave band as electromagnetic

raidiation mwhich is comprised of photons which are flat matter or two dimensional.

A supposedly absurd theory purported that our universe is a computer program supported by physicisits like Dr. Neil deGrasse Tyson of the Hayden Planetarium may seem absurd but if this uniberse is comprised of flat matter which has two macroscopic dimensions and a near infinite number of string sized dimensions which comprise a super-brane or supermembrane as purported in M Theory then this would be compatible with a computer program comprised of a multi-googlplex of bits of data or information forming the program.This is compatible with the dolution to the Hawking Paradox which shows that the bits of information at the evetnt horizon of any active black hole is dispersed or dissolved in space-time as a diffusive process such as a suspension where the bits are so miniscule that they can't be detected by any present technology. What would control the program though?The Higgs Field comprised of tachyons and the massive Higgs Boson and how would the "computer" be turned on and off? Is the Big Bang where a rogue tachyon drops to the speed of light causing a a time oscillation paradox and the vortex of space-time from an infinite number of parallel planes be the turning on by the deterministic rogue tachyon dropping from above light speed to the speed of light causing the time oscillation paradox.And would a Big Crunch be the turning off of the computer program? Many of these ideas would br very difficult to prove but new evidence just released.On the Jpurnal of Physics Review Letters there is evidence that a two dimensional matrix exists with googolplex of microdata which this author believes forms a super-BRANE. It solves problem's with Einstein's Theory of Gravvity as in the two dimensional state the curvature of space-time caused by mass reaches as asymptototic limit approaching zero without reaching it in the two macroscopic dimensions while the superBrane emcompasses the other infinite dimensions.This two dimensional world sheet helps to unify quantum mechanics and trlativity and while the "Computer Program Hypothesis"seems unlikely and ridiculous;it isn't impossible if one can define what the computer ismthe program is and the operator of the computer is..Still it is the opinion of this author that this theory is based on out new technology based on computers and computer software and would have been unheard of in the 1970's or 80's and would have been laughed our of the scientific profession so many Physicisits would still like askance at this theory.

THE TWO DIMENSIONAL UNIVERSE

THE TWO DIMENSIONAL MODEL BASED ON THE YANG BAXTER RELATION IS

BASED ON A PAARTITION FUNCTION OF A TWO DIEMSNIONAL LATTICE FOR

WHICH ELLIPTICAL JOSEPHSONVORTICIES ACT AS FLAT MATTER WITHIN A

VIBRATING MATRIX OF PHOTONS TRAPPED WITHIN BOSOEINSTEINIAN

CONDENSATE AT JUST ABOVE ZERO DEGREES KELVIN.THE EQUATION

$Z=n\{\exp\left[-E(n)/kT\right.$ where E(n)is the energy of the Nth state of matter,K=Boltzman

Constant and T=temperature in degrees kelvin.The Free Energy(F)=-kT ln Z and Z is

defined abpve.The Boltzman equation relates to energy states at phase transitions

as in the latent state of freezing or mLf and of vaporization as mLv.At absolute zero

there is a confluence of vibrating photons trapped in a matrix accounting for the

mass gap in Yang Mills Theory and as mentioned previously in this book.The

correlation functions between spins or isospins can be

lbelled

σi andσj revealing criticality seeking behavior as the metric g ij regarding parameters of crii

Field theory has a !:1 association with the fields of the Ising Model.

initial and j the final event. This expreession is $g_{ij} = <\sigma_i \sigma_j> - <\sigma_i><\sigma_j>$

and depends on the distance separating the states explained by the variable x with regard to tl

Distance and explains "Spooky Action at a Distance" as x approaches infinity as a

distance between the states. $g_{ij} \approx x^{-}$

$$\frac{\tau e^{-x}}{\zeta} \; where \; \zeta =$$

correlation length and τ relates to relative time as x is the distance between states. As the cor

elation length approaches infinity we have a phase transition ind in any phase of flat

matter such as Elliptical Josephson Vorticies the temperature of the system can

vary as there are 252 discreet states of matter in any active black hole. These states

under superpressures included radiation which is flat matter in states such as

liquid,solid BosoEinsteinian Condensate and would neatly fit as Elliptical Josephson

Vorticies fitting into the vortex of space-time as it constricts toward zero as the

event horizon of any active black hole is approached. This is one reason why a

hologram or flat matter is imprinted at the event horizon in terms of microdata and

Field theory has a !:1 association with the fields of the Ising Model.

evaporates or dissolves into space-time as microdate over a 360 degree area as well

as via Hawking Radiation spuming outward from the event horizon.Basically these

data are super-compactified into a two dimensional state where the other

dimensions are super-compactified into a Super-brane while the two macroscopic

dimensions can be viewed like a stop action photograph.The Ising Model has an

energy operator

as $\epsilon n =$

$\sigma n \sigma n +$

1 *at criticality and reveal conformal invariance whereby the relation to conformral*

Field theory has a !:1 association with the fields of the Ising Model.

CHAPTER XXV (SPACE AND BOUNDARIES)

In order to define space one must define boundaries or interfaces. Space by definition is a container without boundaries   with N dimensions where N<infinity and according to String Theory is 26 compactified to 10 or 11 depending on whether super-gravity is included or not. Do contents require a container? You can't have contents without a container as contents cannot exist without a container. Right? Thoughts exist without a container and are contents but do thoughts have mass? K-suryon waves have a specific frequency and energy level therefore thoughts have mass, If thoughts have mass they can be considered contents but to calculate the frequency, wavelength and energy levels of thought(see book s" The Nth Power" and "Megaphysics III; Nothing Doesn't Exist)".Being so the container is the frontal,parietal and temporal lobes of a brain(any brain) diffusing outwardly into space which is infinite. Therefore thoughts require a container and are contents. Conversely a container DOES NOT require contents. Space-time is an infinite container primarily acting as a perfect

fluid. A container does NOT require boundaries and therefore space-time as infinite

by definition has no boundaries (Steven Hawking). If there were boundaries

between space and spacelessness (which is impossible) then there would be an

interface between space and spacelessness and between spacelessness and space.

According to the past and future time cones anything outside of them is "nothing" or

spacelessness however while there may be a boundary between space and

spacelessness there can be NO BOUNDARY between spacelessness and space as a

boundary would have to be included in the null set making spacelessness something

and therefore space as the contents would be the boundary layer and the

spacelessness. Before "The First Event" there were fermions with no mass, tachyons

with a negative mass or positive mass squared and photons which have a mass of

5.1×10^{-19} kV/c^2. Space is infinite as is time and the dimensions are conserved in

the system containing everything but those dimensions do not form until after "The

Time Oscillation Paradox". Prior to that there were D-0-branes without boundaries

due to the Cosmologic Constant being the ground state energy level preventing the

infinite number of planes from touching each other and forming dimensions. Also a true boundary or interface can be subdivided down an infinite number of times with

the space between the interfaces being conserved by the kinetic energy associated

with Dark Energy or The Cosmologic Constant times infinity. Photons always

existed also with entanglement between grouping of photons all having the mass of

5x10^-19kv.c2 and a corresponding energy level from Bohr's Equation

$E=h\upsilon$ $where\ \upsilon = frequency\ of\ the\ photon\ and\ E = energy\ level\ of\ a\ photon.$

Space is a container comprised of leptons and gluons (fermionic stage is the vacuum

stage(Kaku) and gluons keep space intact except during a "Big Crunch" which is a

whirlpool effect of fluid space-time. Both leptons and gluons are virtually massless

but still have miniscule mass which is under the mass of a string   therefore space

has a miniscule mass far under the building blocks of everything or strings

comprising the membranes in M-theory. If gluons were disrupted by an excess of

Dark Energy the fabric of fluid space-time with the Bose Sea enclosed then a

boundary would develop and therefore an interface (worm hole?) Boundaries

within space-time and other dimensions can be described in M theory as

CHAPTER XXVI(THE INFINITE MOMENTUM LIMIT)

There is an equation g
=N/R
∞   *which states that any metric(that which is measurable)equals the number of eigenstates*

of energy divided by the infinite momentum limit. As the metric g can be applied to

open or closed strings with their subcategories being membranes then the energy of

s string can be determined by its eigen-state (which must include phase or state

transformation e.g. from a solid to a liquid to a gas) as determined by the Boltzmann

Equation and Boltzmann Constant. The infinite momentum limit involves the

stress-energy tensor of that metric between gravity , anti-gravity and inertia as well

as electromagnetism ,the Strong Force and the weak force of nuclear decay. When

one goes to the infinite momentum limit which is approached by any active black

hole event horizon the stress-energy tensor approaches infinity as space-time

approaches the infinite curvature of a point with time constricting space toward

zero or a flushing effect or whirlpool effect as the event horizon is approached. The

inertial mass can be huge as the Lorenzian Transformations indicate when the

speed of light is approached(a false boundary)and velocity actually is over

2.99x10^8 meters/sec with a near infinite inertial mass and near infinite curvature

of space-time. The infinite momentum limit is never approached as it is an

asymptote and while the metric g must approach zero it never reaches it (division

by infinity =0) making the metric g approach the mass of a string by the equation

mass)^2=2pi(Tension of a string)(sum of the number of eigen-states of energy

delineated by $\Sigma\ N$) *where sigma is the sum of all the states of strings*. This

emanates from the ground state energy level or the Cosmologic Constant or 10^-55

joules to mu or the maximum free energy of the system incorporating the states. The

mass of the string or SUPERSTRING is the square root of 2piTension (number of

states) or mass= $\sqrt{2\pi\ 10^{39} tons\ (\Sigma_\Lambda^\mu N)}$.   N/R=g ab   where N= number of eigen-

states tons=(2000lbs/ton)(1 kg/2.2

lbs.)=10^39(2x10^3)/2.2=2x10^42/2.2=0.91x10^42 kg which is 60 percent of the

mass of our universe which is 10^54-10^64 kgs.9.1x10^43 kg would comprise

virtually all the mass including dark matter and dark energy (if matter and energy

are inter-changible) to made up virtually all of the galaxies as superstrings or membranes due to the extraordinary potential energy comprising the massive

tension of the string and converting it into the mass equation(which could also

apply to the Higgs Boson or Higgs Field). It is the opinion of this author that for a

string comprised of 10^-33 cm the only conceivable way the tension and mass can

be that great would be if the string was merged with other string s(open or closed)

in such a way that space-time's curvature would be so great that space would be

constricted to a very small partition and time would be significantly dilated or

slowed down such as in "The First Event" just at or before the Time Oscillation

Paradox bridging to immediately after the Time Oscillation Paradox at which point

the earliest strings would be that massive and would curve space-time toward

infinity . The curvature of space-time for 9.1x10^43 kg/string would be 8piG/c4T

ab=N/R where R=infinite momentum tensor regarding stress-energy from the initial

to the final event with N=number of eigen-states of energy. As a result $1/R^2 = \Lambda$ *or*

1/10^-55 j.=R^2 or 10^55)^1/2= 10^28 radians where 2 pi radians is 180 degrees

which mimicks the reciprocal of the total space-time curvature of our present universe which is 5x10^-29 radians. This much potential energy would mimic that

of Planck Time during "The Big Bang" . The location of the site of Cosmic Inflation or

"The Big Bang" is difficult to ascertain as the measuring device and observer is part

of what's being measured. If our universe(which isn't everything) started as a

quantum bubble after a "Big Crunch" of a previous universe or universes the

location of the "Big Bang" or Inflation would be everywhere in our universe if and

only if the explosion or inflation had PERFECT SYMMETRY. IF ONE PORTION OF

THE HYPERROTATING QUANTUM BUBBLE WERE TO INFLATE OR EXPLODATE

PRIOR TO OR AFTER ANOTHER PORTION OF THE QUANTUM BUBBLE there would

be "clumpiness" as was shown in the WOMP diagrams of 1996. There wasn't perfect

symmetry or total isotropism (observational symmetry) and if 50% of the quantum

bubble was matter and 50% anti-matter as proposed there were be a spin causing

the formation of two poles(Megaphysics II,MegaphysicsIII) with a central core of

site of the initial explosion or inflation and if there were a symmetrical throat

leading to a structure analogous to the Robinson Congruence(Megaphysics;A New

Look at the Universe(2003)Dorrance)the orb blast or inflation effect would be perfectly symmetrical and therefore the central core would correspond to a

supermassive BLACK HOLE in our perceived universe with the remainder of the

quantum bubble being the remainder of the universe. This has been borne out

recently showing a confluence of black holes as far out as the light permits

visualization by the Hubble Telescope some 12-13 billion light years away with the

universe being approximately 10^{10} light years in diameter that spot or black hole

would be the center of mass(mass=10^{54}-10^{64} kg) of our universe. It would also

be a point of maximum rotation for our universe causing many ripples in space-time

emanating out from that source although much of the universe still has a curvature

of only 5×10^{-29} radians which is asymptotically flat. So based on these ideas and

data the universe has a center of mass(gravity) and a specific location for "The Big

Bang" or Inflation which was not the first event. Space is a container without

boundaries as boundaries or interfaces have something between the space side of

the interface and the other side of the interface. Without boundaries space is

infinite.

CHAPTER XXVII (27)

THE STRESS ENERGY TENSOR OF MATTER AND HOW IT RELATES TO
EIGENSTATES OF ENERGY

The term "g" is any metric and g=N/R whereby N=eigen-states of energy and R

refers to momentum. Momentum from I to j and j to I (0 to max and max to 0)

relates to the stress energy tensor of matter in such a way that the result is the

ground state energy level or the cosmologic constant. In the previous chapter it was

determined that g=N/R(infinity) as the infinite momentum limit at the event

horizon of a black hole however any momentum is R based on the Stress Energy

Tensor T ab/T ba times the gravitational coupling constant

$8\frac{\pi G}{c4}$. $So\ basically\ g\ ab =$

$N \div 8\pi\ G \div c^4\ \ T\ ab\ \ \ or\ \ \ \ 8\pi G \div \frac{c4}{8\pi}G \div\ \ c4\ or\ 1\ T\ ab\ \div$

$T\ ba\ \ or\ \ \Lambda.\ As\ a\ result\ \ g\ ab =$

$N\ \frac{\text{eigensates of energy}}{\Lambda}.$ .. The result is is any metric $=$

N eigenstates of $\frac{\text{energy}}{}$ $the\ cosmologic\ constant.\ \ Using\ Einsteins's Equation\ 8\pi G\ T\ ab =$

$R\ ab - \frac{1}{2}R\ g\ ab.\ \ Spacetime = \frac{space}{mass}\ equals\ \ space \div i\hbar\rho.$ or g ab=i$\hbar\rho$ ab $=$

$2\pi(space \div spacetime)$ or $ih\rho = space \div spacetime = \frac{1}{time}$ so $g\,ab = ih\rho = 2\pi \div$

$time$. This states that any metric $=$

the square root of -1 times Planck's Constant

Times the energy density of matter or N eigen-states of energy from the cosmologic

constant to mu the highest free energy equaling 2 pi/time.

g
$ab = ih\rho\,ab =$
$2\pi \div time$ as the 2π comes from $\hbar$ going to h in the denominator

and the 2 pi goes into the numerator as it's the denominator of a denominator while

space divided by space-time becomes 1/time which implies that times' arrow

would go backwards as in the case of the tachyon but entropy from the Second Law

of Thermodynamics states that energy goes from a more ordered state to a less

ordered state and our universe is expanding not contracting. The expression of

1/time is not synonymous with – time but rather indicates RECIPROCAL

CURVATURE OF SPACE CAUSED BY THE UNCOILING OF DILATED TIME IN THE

DIRECTION OF CONSTRICTING OR SPEEDING UP. AS TIME SPEEDS UP SPACE

DILATES AND BECOMES CLOSER TO AYMPTOTIC FLATNESS. This occurs with Dark Energy predominating which it does and anti-gravity associated with Dark Energy and the Cosmologic Constant. Based on these premises times' arrow goes backwards with entropy going positive (degree of disorder) and the expansion goes on until Dark Energy breaks all matter into fermions and eventually disrupts space but breaking gluons holding together the leptons in the fermionic or vacuum state causing a Big Crunch which is a whirlpool of space-time going 180 degrees or 2 pi radians from times' arrow going forward. Therefore as black holes follow negative entropy or go from an disordered state to a more ordered state and a black hole causes an expansion of galaxies to reverse into a contraction the reciprocal of time would indicate a monsterous black hole(possible the black hole at the location of "The Big Bang" to have a constant effect of gravity over the anti-gravity effect of Dark Energy curving space-time inwardly as in a "Big Crunch" would also conclude that that Big Crunch has already started and that black holes are interceding throughout our universe reversing times' arrow to point negative(albeit weakly so) and reduce the chaos factor or increasing entropy to a lesser degree of increase. It also implies a disruption of space due to Dark Energy which have myriad non-local

effects throughout our universe(in the multiverse which has reverse effects in other

universes).In all actuality the reciprocal of time means that it DILATES SPACE AS

WITH ANTI-GRAVITY EFFECT rather than constricting space as in the effect of

gravity. As a majority of the energy of our universe is Dark Energy and Dark Energy

fuels the expansion then the dilation of space by the reciprocal of time makes sense

as does the equation $8\pi GT\ ab = \frac{1}{R^2}$ *where R is the curvature of space* −

time and T is the Stress Energy Tensor.

The only way that times' arrow would point backwards would be if the second law

of thermodynamics or entropy would go from an area of increased entropy to one of

decreases entropy and the accelerating expansion of the galaxies away from each

other (shown by red shifts) would slow down and become a contraction or shift to

the ultraviolet. This was discussed in this author's book "What is the Dimension of

Time?"(2014) Authorhouse. This could occur if there was a gargantuan black hole

at the location of "The Big Bang" or Inflation that engulfed everything including all

759 billion galaxies and compressed space-time to a point with infinite curvature

but with a mass so huge it would encompass all the superstrings of this universe at

approximately 10^20-10^22 kg which is approximately forty percent of our

universe with Dark Energy encompassing the other 60 percent and uncoiling the

effect of gravity from a hypercoiled effect to a flattening effect such that the

expansion would occur outside the black hole(60 % of our universe) and the law of

entropy would be positive instead of negative. If the observer is within the

gargantuan black hole and observing mass within that hole times' arrow would

point negative as at the event horizon time would dilate or slow before reversing

inside that black hole. However the probability of this scenario is very remote at

best as surviving the interior of a black hole would be difficult as the temperature

and pressure of that huge mass might be incompatible with intelligent life and

measuring devices according to the weak anthropic principle. As a conclusion times'

arrow would point positive unless there were other undiscovered factors besides

entropy and expansion making the arrow point backwards .Note: Even 1 degree or

1 second of arc backwards would still be backwards and it had been shown that

fermions can travel backwards in time(New Scientist). Again times arrow is always

100% in one direction but may be 67 % forward and 33% backwards or 89%

forward and 11% backwards and the percentage of fermions that travel with times'

arrow in a direction opposite to the observer is measurable at The Hadron Collider

in Cern, Switzerland. However this is unlikely do to the uncoiling effect of time on

space with reciprocal curvature of space .

CHAPTER XXVIII(28) THE FRIEDMANN EQUATIONS

   Space-time is asymptotically flat due to the equation space-time=space/mass or

mass=space/space-time which is mass=1/time. This shows that the fourth

dimension or the dimension of time is asymptotically flat during the accelerating

expansion of all the galaxies away from each other in the near vacuum of space.

Space is curved inward by the effect of gravity and time is dilated or slowed to wear

it acts like a lasso constricting space toward a smaller region as the event horizon of

a black hole is approached and mass greatly increases.. In the near vacuum of deep

space with Dark Energy acting as anti-gravity (the Cosmologic Constant times

infinity until finally braking up Dark Matter into fermions) space is dilated with a lesser and lesser curving effect from time like a lasso uncoiling and approaching a

straight line causing asymptotic flatness. As a result with a positive entropy

universe and a Friedman Type I open flat expanding universe the dimension of time

is clearly inversely proportional to mass such that a small mass reveals a large faster

time or mass=1/time.   The Friedman Equations are based on the volume of a sphere

and the volume of a cone respectively. The volume of a sphere is Volume=4/3 πr^3

and in a perfect sphere after "The Big Bang" the acceleration or Hubble factor would

be a constant which it isn't. This makes inflation a more likely scenario that the Big

Bang giving credence to Alan Guth's theory which also includes the multiverse in

fluid space-time. a:/a={-4/3$\pi(G)(\rho + \frac{3p}{c2})$} + $\Lambda c2/3 - 8/3\pi G\rho$}R^2 = $-kc2$

8
$\frac{8\pi G}{c4}$ *is the gravitational coupling constant or Einstein Constant and this is divided*

by 3 times rho or the energy density of matter from Poisson's E quation. So

8/3(piGrho) and

Λ relates to the accelerated expansion of Dark Energy times the speed of light squared

also divided by 3 times R^2 which relates to the push or energy of expansion

flattening out space-time formulating "-k" signifying anti-gravity times c^2 . As a

result a:/a=-

$$4/3\pi G\left(\rho+\frac{3p}{c2}\right)+\frac{\Lambda c2}{3} \quad as\ a:+\frac{kc2}{a2}=$$

$$8\pi G\rho+\Lambda c2 \quad \div$$

3 *and relates to the volume of a sphere. The accleeratio is time exponenetially =*

$$1/H\sqrt{3c2/8\pi G\rho}$$

Here H is the Hubble Expansion factor and the 3 in the denominator relates

throughout to the volume of an ever expanding uniform sphere.

The volume of a cone which is a Big Crunch followed by a Big Bang or the entrance

and egress of a black hole is Volume=$1/3(\pi r2h)$ where s or surface area of the cone

height (h) radius(r)of the cone is s=$\sqrt{r2+h2}$. S.A.=L=rs with

s=$\sqrt{r2+h2}$ = πrs = *length or surface area.* decreasing or increasing as fluid

space-time (swirling down or spuming up). Cosmic Inflation has changes in

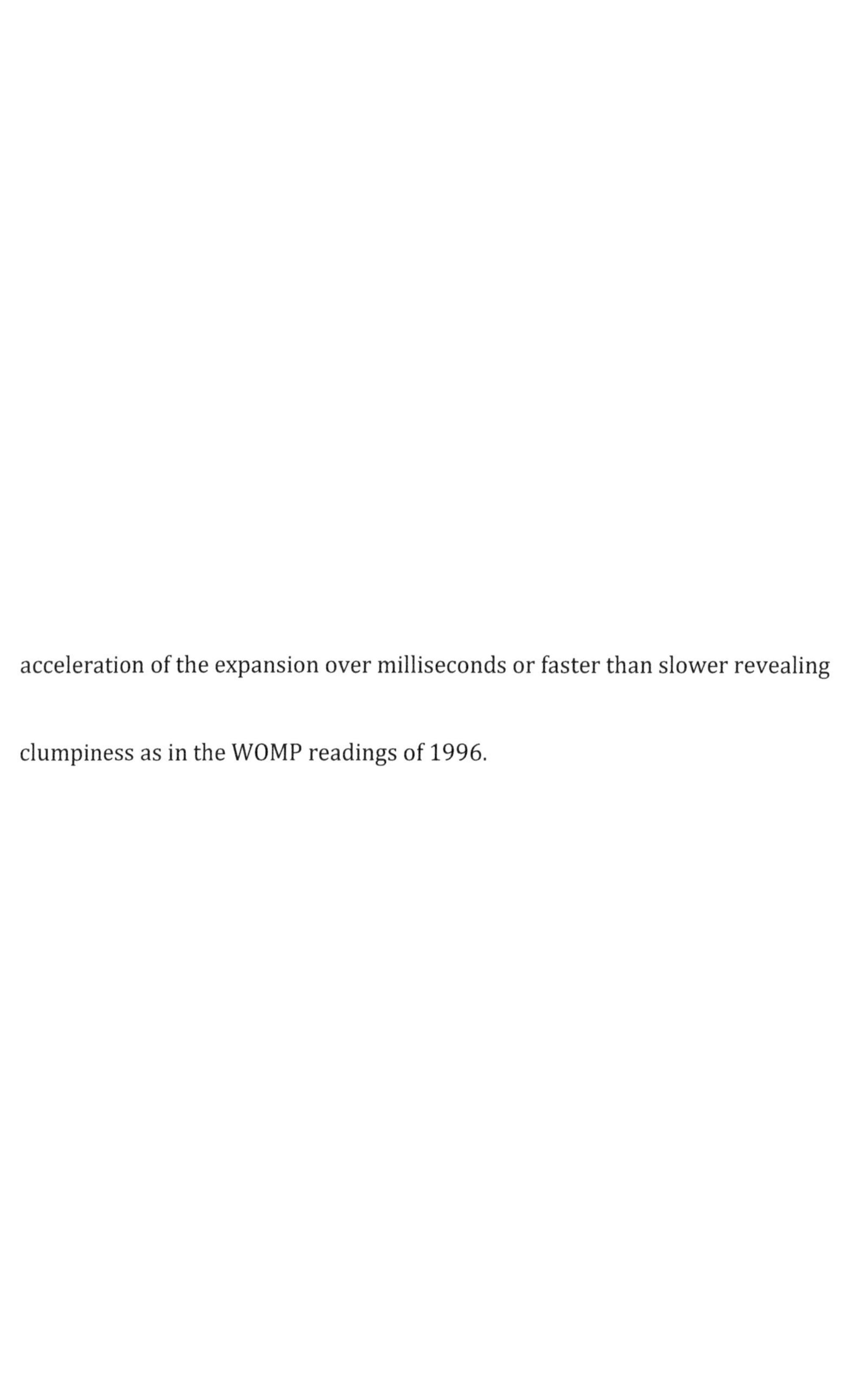

acceleration of the expansion over milliseconds or faster than slower revealing

clumpiness as in the WOMP readings of 1996.

CHAPTER XXVIII

HOW DO GRADIENTS WORK TO ELIMINATE BOUNDARIES?

In order to prove that our universe is not a computer simulation one must rule out

an "on-off" switch and as Steven Hawking stated "There are no boundaries "that are

true boundaries. Instead there are Gradients   which are interfaces which have the

properties of both extremes layers from the inner layer which has properties of the

pre-boundary section to the inner layer which has properties of the outer layer.

Utilizing this concept one can see that the expression space/space-time=1/time

doesn't necessarily imply that times' arrow points in the opposite direction to time

but AS A RECIPROCAL WITH TIME AS BEING A DIMENSION INSTEAD OF TIME

CURVING SPACE INWARD AS IN THE EFFECT OF GRAVITY in the presence of

extreme mass reciprocal time reveals reciprocal curvature of space-time with the

"lasso" of time unwinding over space causing less curvature and approaching

asymptotic flatness as in the near vacuum of deep space with the expansion being

fueled by Dark Energy the cosmologic constant times infinity from the first event.

This must be the case as the likelihood that the second law of thermodynamics showing an increase in entropy as the expansion of fluid space-time fueled by Dark

Energy has one approach an increase in entropy while in a Black Hole there is a

decrease in entropy down to 0.29 as determined by Steven Hawking and his

equation for black hole entropy $S=2\pi\sqrt{NQ1Q5}$ where N are different states

showing that radiation can be liquid under these extreme pressures. Also utilizing M

theory Q1 would be analogous to a 1-brane or the monopole as an electron field

under extreme compression while Q5 would be analogous to the 5-brane

representing 4 space or space-time or hyper-curled space by dilated time.

Conversely with the laws of entropy being intact in an accelerating expansion the

term" 1/time " makes more sense than having a supermassive black hole

permeating all the galaxies in which entropy is decreasing or having the expansion

magically become a contraction (which is the mirror image of what are galaxies are

doing with their red-shifts.

CHAPTER XXXII(32)

A NEW TWIST ON THE EQUATION OF EVERYTHING

THE SCRODINGER EQUATION STATES $i\hbar\frac{\partial\Psi}{dt} = \mathcal{H}\Psi$

Also the position operator x i and the momentum operator p i $=i\hbar$

To include entanglement $\sum_{\Lambda}^{\mu} 1/\sqrt{N}$

Is

i

$i\hbar$ *as the infinite sum of all eigenstates incorporating the position and momentum*

Operators. There fore the anti-derivative of curves space-time is $\Gamma - 1$ $i\hbar$ Ψ/t

as it's the anti-derivative or integral sum of the left side of the Schrodinger Equation

with entanglement being $1/\sqrt{N}$ states or N=ne-E(n)/k T energy states from the

Boltzmann Equation. Therefore $\Gamma - !)abcd$ $i\hbar$ $1/\sqrt{N}$ or $1/\sqrt{E(N)}$ *from* Λ *to* μ

this equals Λ $+$ $or -$ $2\pi c^2$ $+ or -$ $2\pi c2$ $[R\ abc + or - 1/2\ R\ g\ ab \div i\hbar\ \rho\ abc$

Incorporating string theory   the mass of a string
$=2\pi$ *(Tension on a string)(sum of energy states or E(n)where by the total free*

Energy of the massive super-stings is
$\frac{\mu}{2}$ *for the radius of a circle with circumference*   being   the mass squared or the
diameter is μ *or the maximu m free energy of the system.* The energy density
of
matter
 ρ *equals the square root of the tension of each string times the number of states*

Of energy if the 2pi in the numerator and 2 pi in the denominator cancel out.So as a

result the energy states or eigen-states from the Cosmologic Constant as the ground

state to Mu the maximum exicited state causes the energy density of matter to go

from the Cosmologic Constant(energy density of a vacuum) to the energy density of

mu as the diameter whereby the former relates the the open nearly flat expanding

universe with almost flat space-time to the energy density of a supermassive black

hole or just after "The Big Bang" or "The Big Crunch" as the nearly infinite sum of all

energy states between these two limits in our universe. The tension of a string or

supersting is approximately 10^39 tons which when multiplied by the sum of states

(quantum states) is a large number in the denominator for mu the maximum

energy state constricting space-time to a point with time dilated towards infinity.

Also the Cosmologic Constant which is 10^-55 in the denominator causes space-time to be huge as c^2(10^55) R a b c +1/2 R g a b where R g a b is anti-gravity

associated with Dark Energy which is 9x10^16(10^55) or 9x10 ^71 kilometers of

space-time which is almost flat being pushed in every direction except backwards

by Dark Energy which is the Cosmologic Constant times infinity from the infinite

parallel lanes concept of the First Event forming the multi-verse . In that instance

the Cosmologic Constant was the Ground State Energy level keeping those parallel

planes from touch each other to form dimensions. So the square root of 2 pi(tension

on each string)(number of quantum or energy states)=mass of a superstring which

should be 10^54-10^64 kg as the mass of our universe if and only if our universe is

a massive superstring pepped with myriad black holes.

CHAPTER XXXIII(33)

WHAT DOES IT ALL MEAN?

1. There is a beautiful order to the chaos of our universe and the other

    universes(if proven) in the multi-verse of the spiral space-time continuum.

    Life and death are part of a cycle. The Yin and the Yang always fight for

    dominance. Those who purport that "life is meaningless" do not follow the

    idea of the role of life in our universe and our galaxy. If intelligent life is like

    a cell in the body of a living soul living in dilated time with energy being

    formed by the spinning of the central black holes acting as dynamos and

    black holes acting as Mallipigian Tubules for excretion and the Big Bang or

    Inflation is like a birth of our universe utilizing space-time from a previous

    universe as in the "Big Bounce Theory. Also is dark matter intertwines like

    a neural net or nervous system in a body of Higgs Bosons or the Higgs Field

    with tachyons being the brain the role of intelligent life has meaning in the

    maintenance of the whole living universe. Time travel backwards or the

    reversal of times' arrow would have a boundary between the region of the time traveler moving forward in time and the environment moving

backwards in time or visa versa. Unless and until a procedure is discovered

to make one's internal aging and clock move in the same direction as the

environment (if both backwards) the time travel may cause a time

oscillation paradox and possibly be destroyed. We are trapped in the

dimension of time moving in one direction (entropy is positive so it's

forward in our reference frame). We exist.

CHAPTER (5)five a.

THE HUMAN BRAIN

When Pennfield did his experiments in 1958 to surgically treat epilepsy he was able to make people re-live past experiences minute by minute. Touching the right area of the temporal or frontal lobe was able to reproduce experiences as accurate as when they were first experienced. In other words as person's or animal's whole lifetime is locked away in their brain and can be reproduced or re-played that a DVD or VHS recorder in as much detail as the original.

  What does this mean with regard to time travel ,ghosts, and the future? If a person or animal is comatose and "their life flashes in front of them" it's the brain re-playing past experiences and interactions which were stored in the neurons. A ghost siting may be a bleed through of past (Now deceased) individuals it the person own brain and may appear as real and solid as they were when first experienced. If one can relive November 22,1963 or September 11th,2001 one can re-experience those events quite clearly, but can also re-experience March 1st,1999 or May 23,1982 just as clearly. A few people have what's called absolute recall where they can recall any day they experienced going back to infancy and further. Most people cannot recall where they were or what they were doing on August 23,2008 or May7,2006 or December 2,2013 but people with absolute recall can. There are only approximately 8 or 9 documented cases of individuals with absolute recall(one of which is an actress on Taxi whose initials are M.H.) and they are truly gifted but this is not intelligence . Intelligence is the ability to understand and apply abstract principles and those with superior intelligence can understand these principles quite well while the "most intelligent"can come up with the discovery of the preinciples and explain them to other gifted people that can follow and understand them. Again, this is not recall and while recall is not a part of I.Q. without it it is difficult to understand and apply abstract principles on an I.Q. test. In the case of the human or animal brain "dead"relatives can be as alive as the perceiver if the right portion of the brain is stimulated. This explains the past but can future events also be locked up in the human or animal brain? Can one's future illnesses hospitalizations and death be imprinted in one's brain. When it comes to "re-incarnation" the past experiences of the same consciousness in previous lives can be locked up and stored in one's brain and be re-played including talking in a language that that person has never spoken or understood. If our future is locked up in our frontal or temeperal lobes we can find our demises' including how and when if that is desirable; however some inhibitory neurochemical interactions with neurons and recpetors may prevent that until such a time(if ever)that release or death can be a blessing rather than a curse. Would one want to know when a comet or asteroid is going to hit the earth and where? If they knew would they try to prevent it changing the future? If one knew of a fatal traffic accident on a specific date would one take precautions to prevent it? This is why there are probably inhibitory mechanisms within the brain preventing knowledge of future events even in the operating room.

The past can be unlocked but the interactions would have to be the same or the observer totally passive to the experience. The future has so many permutations acting as choices(the human brain can make 28,000 choices each and every day) which involves consciousness and which the human is not in a Pontine Coma suffering from Akinetic Mutism where the body and mind are dissociated similar to thatIf one could "unlock"future event by stimulating certain areas of the brain then when you that making ths stimulators and the individuals being stimulated. If the future is unraveled including one's death at a future time then that future would actually be part of the past. When one encounters the as aesthetic ketamine. one is somewhere else doing something else different than their bodies and that could be part of an unlocked past or future event in the brain.

MEGAPHYSICS II :A DISCOURSE ON HOW SPIRAL SPACETIME LEADS TO THE EQUATION OF EVERYTHING

BY   DR. MITCHELL ALBERT WICK
FORWARD AND INTRODUCTION
  In this author's first book "Mega physics   ,A New Look at the Universe ",it was

postulated that space time curvature followed the spiral fractal formula and was

used to locate the quantum ground state in a Relativistic Universe. This is due to the

rotational component of the open "flat" expanding universe (Friedmann Type II)

and the rotational vector referred to Godel' s Rotating Universe.

  The area of maximum rotation occurred at Planck's Time 10-43 sec and gradually

flattened out with the accelerated expansion after "The Big Bang" until it reached a

curvature equaling the space-time curvature metric of Albert Einstein based on the

Action Formula S=-1/2k2(-g)1/2R where k is the gravitational coupling constant S

is the action of any metric g ,and R is the curvature variant on space-time caused by

the metric g.

  There are two super-symmetric manifolds of space-time as dictated by Quantum

Field Theory and brought to light in this author's second book "The Equation of

Rked on the infinite dimension Hamilton-Jacobi Equations with regard to
viscosity solutions in an article 11.

Everything" as -1/2e-in cot theta   and -1/2 e in cot theta which correspond to l n (u)

and –l n(u) which when superimposed come to flat space time corresponding to

one. One component respponds to clockwise unwinding rotation to "The Big Swirl"

and the other to the counter-clockwise rotation of "The Big Swirl" or anti-Swirl. As

pointed out in "The Equation of Everything" theta is the angle of trajectory of "The

Big Bang" and n is the number of dimensions .e=2.71828 and i=(-1)^1/2.

BACKGROUND

In this author's first book "Mega physics ,A New Look at the Universe "the geometry
of space-time was postulated as spiral. This is due to the combination of expansion
and rotation where if a slice of space-time is made through an expanding and
rotating object is cut and the infinite sum is calculated; it is shown that space-time is
spiral and follows the spiral fractal formula of
$\mathrm{Npi}/2i^{\int \frac{du}{u}}$ *where i is the square root of* $-1.$ *This also equals* $-\frac{1}{2e}-$
incotangent theta as the integral of $\frac{du}{u}$ *is* $\ln u.$ *Over the past eight decades, postulates have beei*

made for the origin of this universe ,including "The Big Bang" and "Inflation

"theories .Also during this time string theory and M Theory have been devised and

evolved as the "New Physics of the 21st Century" .Despite this myriad attempts made

by physicists have failed to show which type of String Theory consistently explains

all phenomena in this universe. This includes the existence of six dimensional

Rked on the infinite dimension Hamilton-Jacobi Equations with regard to   2
viscosity solutions in an article 11.

curled-up manifolds called Calabi -Yau Manifolds. The hybrid of all five string

theories is M theory(Membrane Theory)which look at all five string theories from

different vantage points. An analogy is if five blind men are describing a horse from

different positions relative to the horse where all five are correct. Despite this all

descriptions of the horse are different .As a result all five descriptions are

considered dual to each other as are all five   string theories. Einstein was able to

show that that space-time curvature is a function of mass as a matrix or manifold in

which an added surface or surfaces relate to a fixed number of planes that are either

moving or stationary in a predictable pattern. Space-time manifolds are displaced by

varying degrees of mass and one can visualize a -ball on a trampoline where the

trampoline is displaced by the mass of the -ball   .Only the trampoline is on every

surface of the ball and the ball is curved by everything in the region of the

trampoline while the trampoline is curving or distorting everything around the ball.

Gravity is the curvature of space-time caused by mass. Gravity is conformal with

space-time and the conformations are determined by the mass exerted on space-

time exerted through gravitons and fermions (although spin 2 vector bosons are

Rked on the infinite dimension Hamilton-Jacobi Equations with regard to   3
viscosity solutions in an article 11.

also considered relating to gravity or the curvature of space-time.) Every point in

curved space with reference to time as the fourth dimension is attracted to every

other point as dictated by the mass of the point with reference to all other points   .

Also inertia is the resistance to push (or pull)by the mass ,so every point resists

push or pull as per the inertial (or resting)mass. This can be described by the Ricci

Tensor   used as a description for resting mass.

Space-time can be considered a metric or g as it acts with a vector and scalar

component as conformal to the relationship or inertial mass acting upon it as gravity

R g ab for the inertial mass R ab. The definition of a manifold based on surfaces

added to a three dimensional surface has ,in the past been related or restricted to

three or greater dimensions .However ,with the advent of flat matter and, elliptical

Josephson vorticies, noted as one or two dimensional if dilated time is considered

and is congruous to any matter that is superheated ,then the definition of a manifold

must be changed to include all dimensions of one or greater ,the first dimension

describes time and the second describes a closed flat string. Einstein predicted that

space-time is flat in the absence of mass and the geodesics of a space-time metric g

Rked on the infinite dimension Hamilton-Jacobi Equations with regard to   4
viscosity solutions in an article 11.

ab has been described as the Lorentzian metric in which it interacts as another

metric. When the expression of inertial mass(push)and gravitational curvature of

space-time caused by the inertial mass where the mass is described as a metric on

the remainder of the space-time manifold are equal ,the scalar of space-time

manifolds is zero and the manifold is flat(zero curvature).

It has been determined by this author that Space-time is directly proportional to

space and inversely proportional to mass.The dimension of perceived time slows

down as the object measured approaches a heavy mass.At the event horizon of any

black hole space-time shrinks towards zero without actually reaching it as an

asymptotic limit. Also space-time is directly proportional to space as in this universe

or space-time manifold time's arrow points forward not backwards .Therefore

space-time=space/mass x constant and the constant is 1/c2 as space/mc2 where

mc2 is the Grand Unification Energy of 10 19 Giga electronvolts is the equation of

"The Big Bang" at Planck Time(10-43 seconds).when space-time is -1/2 e –in

cotangent theta where n=number of dimensions and the dimensions approach

Rked on the infinite dimension Hamilton-Jacobi Equations with regard to
viscosity solutions in an article 11.

infinity as an asymptotic function ;the expression for space-time approaches zero

without reaching it as in the event horizon of a black hole. This is the quantum

ground state in a Relativistic Universe and unifies Quantum Mechanics and

Relativity.

CHAPTER ONE

Rked on the infinite dimension Hamilton-Jacobi Equations with regard to
viscosity solutions in an article 11.

BACKGROUND AND STRING THEORY

Strings are the smallest units postulated at Planck Length which is 10-33cm.

These strings are basically two dimensional units of either energy or matter

depending on which source you read and move in multiple planes. Strings can twist

turn ,rotate and connect or disconnect with other strings. It is postulated that are

made of two types ;open and closed. They move in 26 dimensions which are

compactified (rolled up and curled)into either 10 or 11 depending on whether the

reader incorporates Supergravity in Quantum Field Theory incorporating gauge

symmetry groups as illustrating the 11th dimension .Space-time is described in units

called orbifolds; which are manifolds or surfaces which twist and turn in the

configuration of a modified cone. A closed string represents a graviton particle or

gravitational movements   which are mimicked or represented by a spin 2 boson.

The world or universe   can be represented as a two dimensional sheet(The World

Sheet)of either closed or open flat strings that vibrate and rotate with reference to

themselves and each other in different combinations which represent twisted or

torsed donuts or toruses in flat combinations which can project into multiple planes

Rked on the infinite dimension Hamilton-Jacobi Equations with regard to   7
viscosity solutions in an article 11.

forming six dimensional twisted or puckered Calabi Yau manifolds. The two

dimensional world sheet mapped topologically with conformal mapping must apply

the rules of symmetry as well as quantum vibration and have to be dealt with even

in a relative vacuum even if this vibration self-annihilated instantly.

Because of the inability to empirically measure string activity   ,some of the

physics community looked askance at string theorists. Despite this   ,physicists such

as Brian Green are trying to quantify string theory with respect to 'The Big Bang' as

it was initially broached by Edwin Hubble or 'The Inflation Theory 'as postulated by

Alan Guth. Both theories were ex nihilo or "out of nothing" and were postulated as

occurring at approximately 13.7 billion years ago although one must differentiate

between Hubble Time and Conformal Time as differentiated by Einstein. Both

theories were tied in with the expanding universe as discovered by Edwin Hubble in

the late 1920's.

String Theory was originally purported in the 1980's when it was discovered

mathematically that nature follows harmonics of musical notes. These harmonics

were incorporated into two dimensional energy components of matter called

Rked on the infinite dimension Hamilton-Jacobi Equations with regard to

viscosity solutions in an article 11.

strings. Closed strings are continuous and formed loops ,double tori, torus

configurations(as mentioned previously),triangles, rectangles and a myriad of other

configurations based on the energy of the string with regard to other strings in

space-time. As mentioned previously the unit of space-time or geodesic is called an

orbifold which was defined earlier. Again this unit of space-time is a manifold or

surface that twists and turns in myriad configurations as do closed and open strings

.Calabi Yau manifolds are derived configurations of orbifolds and are generally

Planck Length or 10-33 cm. Open strings were coined by Dr .Roger Penrose as

twisters another description of closed and open strings. Twister theory incorporates

imaginary numbers with dimensions to indicate matter in a shadow universe

,similar but mathematically different from this's author's -1/2e-incotangent theta

incorporated into space-time with regards to quantum mechanics. At Planck Length

space-time is curved in and around itself like mini-black holes giving it infinite

curvature and transforming it into quantum foam from continuous space-time

according to quantum theory while Einstein purported that space-time was

continuous down to an infinitely small size. Also ,Einstein stated that what was what

Rked on the infinite dimension Hamilton-Jacobi Equations with regard to   9
viscosity solutions in an article 11.

appeared to be the force of gravity was in fact space-time curvature caused by mass.

This is why R g ab is the space-time curvature metric which describes the effect of

graviton space-time R causing curvature . String frequencies are based on the note

frequency f=1/2L(T/m)1/2 where L is the length of the string(guitar string not

string which is smallest unit of matter) (T is the tension and m is the mass of the

string .Decrease the length of a guitar string and the frequency increases .Place a

finger at the midpoint of the string and the frequency splits in two. These concepts

are the basis of string theory's correlations with harmonics or as Pythagorus stated

"the music of the spheres" which was first thought of by the Greeks. Open strings

split and join and closed strings mimic the spin 2 vector bosons(as previously

mentioned)which are quantized units of gravity or mass acting on curving space-

time geodesics .The spin 2 vector boson curve space-time as the carrier particles for

each and every target mass with the Ricci Tensor R ab. describing the resting mass

of the metric g ab. Again this is the effect of gravity rather than it being a force

according to Einstein .Hadrons were explained as strongly interacting elementary

particles. These spin 2 vector bosons appear as closed loop strings and as a basic

Rked on the infinite dimension Hamilton-Jacobi Equations with regard to

viscosity solutions in an article 11.

building block of matter spin 2 vector boson pervades all mass everywhere and curve space-time as that mass Dimensionality of the space in which strings vibrate relate to1-(D-2)/24which will only follow relativistic covariance if string moves in a net of 0 dimensional space where a total of 26 non-compactified dimensions exist in Quantum Field Theory and with the SO(32)gauge symmetry group this problem was cleared up partially.C.Everett Peat p.99ClosedStrings appeared in type II string theory like massless bosons with a spin of 2.This relates to the spin2 vector boson involving the effect of gravity. Are strings flat matter or energy? The answer is probably both.One could refer to the Higgs Boson which was coined "The God Particle" which would bear out the boson as being the fundamental building block of matter and energy all pervading .As previously mentioned there are five disparate string theories which have duality to each other .They are type I ,type II ,type IIa, Heterotic 8x8 and the SO(32) string theories which when observed all as one unit forms Superstrings are groupoids of strings with the mathematical equations that describes the way a sting moves and vibrates ,making sure they follow the rules of relativity ,quantizing the equations of the relativistic string ,and making sure that all Rked on the infinite dimension Hamilton-Jacobi Equations with regard to viscosity solutions in an article 11.

strings are supersymmetric and follow the rules of gauge symmetry with strings

associated with gauge symmetry groups involving elementary particles .Quantum

Field Theory involves covariant systems where an action can move from one

coordinate system to another as in parity involved with the CPT Theorem which

involves covariance. The tension of a string was determined as 10^39tons Peat

p.101 making it appear to be more energy than matter with a miniscule mass

Although the m^2 operator for

$2\pi T$ for the open and closed strings determine it has a mass. $m^2 = 2\pi T \Sigma n =$

$1\infty\Sigma = i = 1\ to\ D - 2\alpha - n^{1}\alpha n^{1}w\ here 2\pi T =$

$\frac{1}{Regge}$ Slope which is $\alpha^{1.2\pi T}$ is a hige number but it's reciprocal is a small number..

Supersymmetry involves properties of iso-spin and strangeness of subatomic

particles . M Theory which is short for Matrix Theory or Membrane Theory

depending on the source Heterotic strings share two dimensions in one;one rotating

clockwise the other counterclockwise with regard to a rotating orbifold or Calabi

Yau Manifold..M theory contains supergravity in the 11th dimension(D=11)n the low

Rked on the infinite dimension Hamilton-Jacobi Equations with regard to   12
viscosity solutions in an article 11.

energy limit and reduces to the type IIa string theory which when compactified

forms a sphere.The infinite momentum limit of the D-0 brane may relate to the

U(N)super Yang Mills TheoryThe D-0 brane or membrane relates to the 10 which

broke up or cleaved into a dimensional state which has a limit of N approaching

infinity for 10 dimensional 0-branes.Based on this argument there were an infinite

number of dimensions in the vacuum pre Big Bang state and the 0-branes were the

building blocks of everything which were cleaved or broken into a six dimensional

component and a four dimensional component where the former is the Ca;abi Yau

Manifold and the latter is four dimensional space-time leading to the 10

compactified dimensions of Type IIa string theory. The idea is consistant with the

Law of Conservation of Dimensions which states that the sum total number of

dimensions of a system in any form is a constant. Infinite momentum(p) relates to

the Unified Energy with super gravity in the 11th dimension as described in

YangMIlls U(N).To be compactified into a circle or sphere type IIa string theory

would indicate world sheet which was two dimensional but spherical or circular in

two dimensions. Closed strings would fit better in a world sheet which was

Rked on the infinite dimension Hamilton-Jacobi Equations with regard to
viscosity solutions in an article 11.

compactified to a circle and this would eliminate duality as the five string theories

would curl up the world sheet to a point with infinite space-time curvature or a

circle of quantum foam when the environment is below Planck Length or 10-33 cm.

CHAPTER TWO

WHAT IS SPACETIME?

Space –time is a term that Albert Einstein coined for   what he considered four

dimensions of length   ,width, height, and time. It was based on the Line Element

ds2=dx2+dy2+dz2-c2dt2+dr2 where r=space-time curvature metric described by

the tensor R g ab. Ds2 is a description of the four coordinate system with regard to

space-time curvature and the relativistic effect of c2dt2. R g ab or r is determined by

R ab or the Ricci Tensor describing inertial mass of an object doing the curving and

the curving is done by the spin 2vector bosons and possibly gravitons or fermions .

Spiral space-time has a k=-I to the n cotangent theta power as suggested by Sir

Roger Penrose and proposed by this author. Flat space-time is space-time without

any curvature and occurs in a vacuum state. Considering the fact that photons have

a measurable mass while in a moving state moving photons can also curve space-

Rked on the infinite dimension Hamilton-Jacobi Equations with regard to   14
viscosity solutions in an article 11.

time as energy c ,however any mass at Planck Length or greater will curve space-

time. It is unknown if mass under Planck Length will curve space-time as there may

be a lower limit of Planck Length for space-time to appear in anything instead of

quantum foam or quantum dots which may be uniform or curved inward or

outward depending on the activity of string components. This quantum foam would

make up the orbifold unit of space-time.

Any mass curves space-time from string sized to that supermassive black hole which

centers each and everyone of the 750 billion galaxies in this universe and can be

displaced by the spiral or cone shaped designation of space-time as it approaches

the event horizon of a black hole where perceived time appears to the intelligent

observer to shrink along with local space to a point which may also be string sized

or 10-33 cm at the center of the black hole although there is no proof that the center

of a black hole isn't greater in size than that of a string.

Einstein's equation of Relativistic Gravity described the Einstein Tensor G ab=R ab-

1/2R g ab=8(pi)T ab where the Einstein Tensor=Ricci Tensor-1/2(Relativistic

Rked on the infinite dimension Hamilton-Jacobi Equations with regard to 15
viscosity solutions in an article 11.

Gravity)=8(pi)T where T=stress energy tensor of the metric g ab. Relativistic Gravity

is the curvature of space-time caused by the metric g ab and the Ricci Tensor

represents the inertial mass of the metric g ab. Inertia-gravity =1/2 but with the

metric tensors being abelian and anti-symmetric Inertia-gravity and gravity-inertia

are equal but opposite in magnitude and direction netting out zero with is the value

of the Einstein Tensor revealing a stress energy tensor of approximately zero.

The equation was derived by the Lorenzian Transformatios but basically says stress

energy is zero in the "Pre-Big Bang "or "Post Big Crunch "epoch .In black holes the

gravitational effect from a collapsed neutron star or galaxy is so great even light

can't escape. Beyond the event horizon of a black hole space-time is collapsed in a

spiral or vortex configuration to almost zero(as mentioned previously)and mass is

collapsed to almost infinite density. This mimicks the pre-Big Bang where the

Einstein Tensor G ab approaches zero.This is illustrated in the tensor expression of

The Equation of Everything where R abcd=R abc-1/2 R g ab/R ab where space-time

is curved inward by the collapsed mass of a black hole.In essence,the space-time

curvature metric is gravity acting on R,which is the spa ce-time that is curved by the

Rked on the infinite dimension Hamilton-Jacobi Equations with regard to 16
viscosity solutions in an article 11.

metric g ab. In the "Pre-Big Bang" epoch(up to 10-43 seconds or Planck Time)the

space-time curvature metric approaches infinite curvature as a point which has

infinite curvature and the extreme mass of the pre-Big Bang quantum bubble curves

extremely small space-time toward infinity without reaching it.In this case gravity

approaches half of a very large non-infinite number and anti-gravity from

antimatter approaches a very large non-infinite number but inertia equals the sum

total of the gravitational and antigravitational metric acting on space-time with

almost infinite curvature. This all indicates that Gravity isn't a force but the effect of

space-time curvature caused by any mass. The action of any metric in s[ace-time R

with the curvature metric of (-g)1/2. R can be described in terms of dimensions

such as d to the nth dimensional power and is extremely useful in string theory with

its postulated 26 dimensions compactified(curled up)to 10.As previously mentioned

Relativistic Gravity is described by 8(pi)Tab=R ab-1/2R g ab where R g ab describes

Relativistic Gravity and its effect on space-time. The space-time curvature metric is

described in the tensor equation R a b c d=R a b c -1/2r g a b/R a b where R a b c d

describes curved Lorenzian space-time R a b c describes flat Mintkowski or

Rked on the infinite dimension Hamilton-Jacobi Equations with regard to   17
viscosity solutions in an article 11.

Riemannian space and R g ab describes the space-time curvature metric acting upon

R a b c causing the curvature caused by the mass described in the Ricci Tensor R ab

describing the inertial mass of metric g ab.The entire expression is multiplied by

1/c2 to give the Relativistic form of the Equation of Everything .Utilizing Einstein's

equation of Relativistic Gravity and this author's equation of everything and solving

the stress energy tensor T ab one gets the gravitational constant G6.67x10-11

newtonmeters/sec2.Setting R ab from both equations equal to each other by the

transitivity postulate a=b b=c therefore a=c   Tij==T ab.T ba where i=initial event

and j=final event and mass=energy/c2 as T i j=

 T ab. Tba=Tij/||c2||2 and 8 pi T=R ab-1/2 R g ab via Einstein's formula where
8(pi)T ab.T ba/c4 reflects the stress energy tensor T ij=R ab-1/2R g ab at zero stress
energy.In other words the Einstein Tensor G ab=T ab.T ba=G(the gravitational
constant)as 8(pi)G/c4 is the gravitational coupling constant k and the stress energy
of zero has constraints of the pre-Big Bang or post-Big Crunch and at event horizons
of black holes. Stress energy is approached by the gravitational constant G=6.67x10-
11 newton meters/sec2 which approaches zero at the point of the Pre Big Bang"
epoch or where the Einstein Tensor approaches zero.The difference is attributed to
weak perturbations as described by the action formula S or the Hamiltonian
Operator for n eigenstates of energy as the 0 or null eigenstate is approached .An
operator is a complex function which operates on another function such as the
LaPlacean Operator ,which acts as a second degree differential equation of the
function it is operating onto the upward limit of the operator. An eigen-state is a
state of matter or energy with regard to the Operator that operates on another
function such as the quantum level of matter with regard to energy.
CHAPTER THREE
THE BIG BANG   ;WHAT WAS IT?

Rked on the infinite dimension Hamilton-Jacobi Equations with regard to   18
viscosity solutions in an article 11.

OVER 13.7 BILLION YEARS AGO a quantum bubble with a 50:50 mix of matter and antimatter underwent either a 360 degree orb blast ,inflation ,or a two phase swirl which continuously expanded with progressive decrease in rotation. Prior to that is up to speculation. There is one theory that two membranes from other universes collided or touched triggering "The Big Bang' or a series of zero-BRANES the building blocks of matter underwent a dimensional recombination like "the popped bedsheet"hypothesis of Dr.Michio Kaku 1 forming a six dimensional manifold in a Calabi Yau configuration and the macroscopic four dimensional manifold of space-time. Another hypothesis is that a previous universe underwent a "Big Crunch" causing time's arrow to reverse and turning positive time into negative time(backwards)due to the implosion until it reached just before time zero when time advanced to Planck Time 10-43 seconds and the matter-antimatter mix exploded. The gravitational effect of antiparticles and particles may have caused "The Big Bang"and resulted in Dark Energy which propelled the accelerated expansion of galaxies away from each other that Edwin Hubble discovered in the late 1920's. According to the math antiparticles repel each other causing such an

Rked on the infinite dimension Hamilton-Jacobi Equations with regard to 19
viscosity solutions in an article 11.

explosive force in just under 50 percent of the matter antimatter mix that from a

Planck Length string sized quantum bubble the"Big Bang" results in a massive

antigravity surge from antiparticle antiparticle repulsion . This phenomenon not

only explains the existence of Dark Energy with pushes galaxies apart from each

other but also the paucity and lack of cohesion of antiparticles in the universe.

Antimatter was first discovered in 1932 and since then myriad antiparticles have

been discovered. Antiparticles are being isolated in Cern,Switzerland in the Hadron

Collider but as of this date the mass of any. antiparticle has not been conclusively

discovered. It has been noted that antiparticles such as the positron have the

opposite charge as matter particles but it has not been determined that antiparticles

have positive gravity. There are some experiments that do show repulsion of

antiparticles in positive gravity fields 2,but to collide particles and antiparticles can

cause annihilation of the particles with the expulsion of energy but this doesn't

necessarily mean that particles and antiparticles attract each other by gravity;only

possibly by charge and electromagnetism which are stronger forces than gravity

Rked on the infinite dimension Hamilton-Jacobi Equations with regard to   20
viscosity solutions in an article 11.

which acts as a weak force but isn't actually a force as previously mentioned. As

antiparticles self repel the energy required to push antiparticles toward each other

would be considerable and might destroy the antiparticles. If antiparticles mutually

repel with antigravity instead of attract with gravity a plausible mechanism for "The

Big Bang"can be made from the quantum bubble. At Planck Time 10-43 seconds a

50:50 mix of matter and antimatter caused an orb blast with a trajectory of theta in

the spacetime equation -1/2e to the i n cotangent theta power where i=the square

root of -1. And no is the number of dimensions either 4 for macroscopic spacetime

or 10 compactified dimensions including Calabi Yau Manifolds or Oribfolds in string

theory. The 2 pi radian orb blast with the mutually repulsive force of antiparticles

forcing an explosion converting over 99.999 per cent of the antimatter into Dark

Energy by the formula E=mc2 postulated by Albert Einstein where m=mass of the

antimatter. Whether the mass is positive or negative is up to speculation but the

Law of Conservation of Energy was purportedly been violated by the "Big

Bang"jnless the potential energy of the quantum bubble equaled the kinetic

energy,heat,and Dark Energy after "The Big Bang ".However the potential energy of

Rked on the infinite dimension Hamilton-Jacobi Equations with regard to

viscosity solutions in an article 11.

the quantum bubble didn't spontaneously appear "ex nihilo" or out of nothing and

may have been from a collision of a matter universe ,antimatter universe(or

membrane) and a relative vacuum forcing the pre-Big Bang implosion which drove

time's arrow backwards and reducing entropy from two universes to a quantum

bubble(making the occurrence a singularity as the Second Law of Thermodynamics

and Time's Arrow pointing forward must be suspended for this acr)As antiparticles

repel dark energy pushed outward in all directions carrying the balance of matter

with which it is subsequently attracted to other matter by gravity but not to

antimatter(which appears to have a lack of cohesion);the formula R j I k l-Rjl

kl(where kl is a superscript to the R in the second term while in the first R j I k l are

subheadings all to indicate covariant and contravariant tensors respectively=-

8(pi){Ge ji}}= g ji in four dimensions of space-time(an opposite curvature or

reciprocal curvature to gravity being antigravity as the sign for—(8 pi{[Ge j i]}=g j

iis negative.The rightside of the equation is the mutually repulsive force of

antiparticles and g j I is the metric of the antiparticle where i=initial event and

j=final event and-8(pi)G where G=the gravitational constant is from the Cosmologic
Rked on the infinite dimension Hamilton-Jacobi Equations with regard to   22
viscosity solutions in an article 11.

Constant ^which reflects the mutually repulsive force pushing galaxies apart

purported due to Dark Energy.E=2.71828 and e j I is the vector product of e from j to

i. This assumes a 360 degree or 2 Pi radian orb blast in the "Big Bang" and R is the

anti-gravitational effect on particles while conversely R j i is the antigravitational

effect on antiparticles. In a 360 degree or two pi radian orb blast the angle of

trajectory is pi radians or 180 degrees. The cosine of pi radians =-1 which explains

the -8(pi)G on the right side of the equation. There are approaching an infinite

number of 180 degree slices in a perfect sphere so the angle of trajectory for an

isotropic universe must be 180 degrees or pi radians.

THE C.P.T. THEOREM   AND ITS' INNATE SYMMETRY OF NATURE

Charge, parity and time whether reversed or not have innate symmetry in

nonlocal systems according to Quantum Field Theory. This indicates that if charge

were reversed as in the positron vs.the electron the magnitude of the charge would

be essentially unchanged. If time were reversed that

$$\Im\Psi(x,t)\Im - 1 = e\ iT\Psi(x,-t) where\ i =$$

Rked on the infinite dimension Hamilton-Jacobi Equations with regard to   23
viscosity solutions in an article 11.

$\sqrt{} - 1$ *and e is to the $i\Phi$ power, the transformation of t to $-$*

t can be shown as commutative as the operator $\Im$ is antiunitary. The operator cancommute wit

The Hamiltonian and still reverse the sign of t.3 With parity X can be replaced with –
X and still have a commutative Hamiltonian Operator such that $(CPT)\mathcal{H}(CPT) - 1 =$
$\mathcal{H}(-x)$. *Consider space $-$*
time curvature of matter and antimatter. Combining matter and antimatter have space $-$
time curvatures which would complement each other cancelling each other out resulting in fla

Space-time when matter and antimatter annihilate each other. This causes

energy=mass of the antiparticle+mass of particleXc2 resulting in the interfitting of

the reciprocal curvatures of space-time for identical particles and antiparticles

resulting in asymptotic flatness. In an antimatter universe of manifold pre-

dominantly anti-matter anti-gravity would attract rather than repel repel anti-

particles due to Parity replacing X by –X with the commutative Hamiltonian

Operator and in this case gravity would repel rather than attract particles for the

same reason. Conversely in a matter dominated universe anti-matter would repel

anti-particles with anti-gravity and matter would attract particles with gravity.

Rked on the infinite dimension Hamilton-Jacobi Equations with regard to 24
viscosity solutions in an article 11.

Outside of weak perturbations the symmetry of nonlocal systems is upheld with the

C.P.T. Theorem.

CHAPTER FOUR

TIME'S ARROW AND THE LAW OF ENTROPY

Time's Arrow states that time will be move forward in our space=time

continuum. The Law of Entropy or the Second Law of Thermodynamics states that

every system or subsystem will always go from a more ordered state to a less

ordered state. This obviously occurs in an open flat expanding universe ,but what

happens in the event that there is a "Big Crunch"? In an implosion   where a less

ordered state goes toward a quantum bubble where space-time develops a rip or

tear space=time can "pop" like a balloon and the second law of thermodynamics

may be violated. Mathematically,   the equation space-time=space/mass(1/c2)

negative space-time equaling negative space/positive mass x 1/c2 which dictates

the rate of compaction of space-time in a "Big Crunch". Also according to Einstein's

Law of Relativistic Gravity   G a b=R a b-1/2R g ab where G a b=0 in a pre-Big Bang

or post Big Crunch epoch. Here R a b stays positive for the Ricci Tensor or inertial

Rked on the infinite dimension Hamilton-Jacobi Equations with regard to   25
viscosity solutions in an article 11.

mass but the space-time curvature metric known as gravity changes sign or the direction of the vectors of the positive mass from $-R\,g\,ab$ to $+R\,g\,ab$ and the stress energy tensor $8(pi)T\,a\,b$ goes equal in magnitude but opposite in direction .Due to the fact that space-time=space-time and with the Bianchi Identity $-1/2R\,g\,a\,b=1/2\,R\,g\,ab$ in magnitude but opposite in direction but since anti-symmetric they net out to zero so the inertial mass of this universe stays the same as the Ricci Tensor $R\,ab$.

Since the only way to prove these conclusions is in a Big Crunch which could occur in an accelerating rate toward Planck Time 10-43 or slowly where the only proof would be a shift toward the ultra-violet with regard to the decelerating expansion of galaxies as measured on the Hubble telescope. As approaching a heavy mass time slows down the mass/per space decreases as in a Black Hole time must slow down in a Big Crunch toward a quantum bubble and in that case "Time's Arrow" would decrease. The equation $\mathbb{R}\,a\,b\,c\,d=R\,a\,b\,c\,-1/2R\,g\,ab/R\,a\,b$ changes to $\mathbb{R}a\,b\,c\,d=R\,a\,b\,c+1/2R\,g\,ab/R\,a\,b$ in the event of a Big Crunch as space-time undergoes reciprocal curvature from the space-time curvature metric called gravity in other word space-time would curve inward in the presence of inertial mass instead of outward and

Rked on the infinite dimension Hamilton-Jacobi Equations with regard to viscosity solutions in an article 11.

with space-time reducing in size to a point or quantum bubble such as is analogous

to a black hole event horizon R a b c+1/2 R g a b is greater than R a b c. The entire

expression R a b c+1/2 R g a b is greater than R a b c d which is decreasing and as R

g ab=-R g ab as anti symmetric tensors then R a b c d=-1(R a b c-1/2R g ab/R a b)

whuch indicates the decreasing direction of time's arrow with the curved Lorenzian

or Riemannian Space-time ℝ a b c d.

BLACK HOLE ENTROPY IS BASED ON STEVEN HAWKING'S FORMULA
S=2(pi)(NQ1Q5)1/2 where s= entropy N is the number of states or eigenstates in a
black hole postulated as 252 separate states and Q1and Q5 are the differential
charge between the first and fifth eigenstates.S=degree of disorder. The one brane
in M theory(membrane theory)is described with the monopole fixed negative
charge suggestive of an electron or positron with the antiparticle and the five-brane
represents 4 –space or curved Lorenzian Space-time which spiral into a black hole
event horizon .As a result the energy of a monopole acting on curved Lorenzian
Space-time across 252 states of matter reveal black hole entropy .Also S(black
hole)=Area/4 Length2 P=c3A/4Gh where h=Planck's Constant P=Planck Length of
10-33 mA=cross sectional area and G is the Gravitational Constant of 6.67 x10-11
newton meters/sec2 as c=3x10 8 meters /second or the speed of light which is a
true boundary for any mass as space-time shrinks to approach zero at that
boundary. Dr .Hawking postulated the entropy of a black hole to be 0.29 which
approaches zero. This indicate that as the cross sectional area approaches 0 or
P(Planck Length)the entropy of a black hole approaches zero(0)according to the
Bekenstein-Hawking Equation .This mimics the entropy(S) of the Quantum Bubble
at pre-Planck Time before "The Big Bang "where a 50:50%0 mix of matter and
antimatter are solidified by enormous pressure into what might be called a lattice
formation much as a diamond would occur. Indeed the central locus of a black hole
post event horizon ight have the same or similar configuration as with tremendous
pressures strange matter and liquid states of matter which would under other
ambient conditions not be liquid or solid. In a "Big Crunch" which may occur in
Planck's Time(10-43 sec)which is fast or a slow leak of space-time like a deflating
balloon would eventually approach 252 eigenstates of a quantum bubble similar to
that of a black hole and measured time would slow down as a shift toward the
ultraviolet would occur as galaxies recede instead of expand. A nuclear clock might

Rked on the infinite dimension Hamilton-Jacobi Equations with regard to   27
viscosity solutions in an article 11.

lose 10-3 seconds for each month that the galaxies recede instead of expanding and would be for most in perceptible except with scientific measurement. Also there would be no clear indication as to when the tail end of a "Big Crunch" would occur in the last 10-43 seconds where everything would shrink to approximately 10-33 cm like the initial quantum bubble. As a result it is extremely difficult to empirically prove that" Time's Arrow" is reversed in a "Big Crunch" also because the energy to reverse the sequencing of events in a universe would require the energy of a Big Crunch.

An alternative explanation for the end of this universe would be "Heat Death "where all matter and antimatter would slow down its' expansion until it eventually stops all fusion and fission in stars would eventually slow down and stop and everything would slow down to a crawl at 2.74 degrees kelvin(the temperature of the background microwave radiation from the "Big Bang")and all matter would "freeze".

CHAPTER FIVE;
WHAT IS SCHWARZCHILD SPACE-TIME AND HOW DOES IT RELATE TO BLACK HOLES?
It is well known that as the event horizon of a black hole is approached space-time approaches zero, time is dilated toward infinity(infinitely long)and mass increases dramatically along with gravity as space-time approaches a point string sized or 10-33cm in de Sitter or anti-de-sitter space. Quasars develop and spume out matter energy(Hawking Radiation)4 and information from the event horizon of a black hole whose origins or poles depend on the location, velocity and mass of the observer and are based on observational viewpoints.

The entropy of a black hole was already discussed in the previous chapter and again was postulated by Hawking as 0.29 where at least 252 different states of matter appear as N in the equation S=2(pi){NQ1Q5)1/2 where Q1 is the membrane boundary of the monopole(as mentioned previously)representing an electron cloud bounded by the membrane known as the 1-BRANE in terms of M Theory and Q5 represents Lorenzian Curved space-time or 4 space on which the 1-brane binds it with the electron(or positron)cloud acting upon it across the 252 states of matter.Q1 is 1.602x10-19 coulombs or the charge of an electron and a unit of space-time referred again as the oribifold and is generally bounded by Planck Length(10-33)cm. Therefore black hole entropy=2(pi)(252x10-19coulombs)(10-33 cm)to the one half power or 2(pi)(4.032x10-54 to the one half power or 2(pi)(4.032x10-108) Which approaches zero entropy. At the point at the event horizon the conformation is spherical and two dimensional with regard to position and action of the observer and the observer's motion. The mass of the information at the Event Horizon relates to its' spherical radius(which decreases toward zero)in Region II of collapsing matter using Schwarzchild Space-time. The event horizon is basically a cork which is a spherical region with extremely dense mass froma collapsed neutron star or galaxy and the hole of the event horizon is basically clogged up by the mass.In a way that matter queues up in that spherical region of collapsing space-time that approaches zero or string sized.Region 1 of Schwarzchild Space-time has a constant

Rked on the infinite dimension Hamilton-Jacobi Equations with regard to viscosity solutions in an article 11.

radius and constant time.Region 2 has a radius approaching twice the mass and time approaching infinite dilation. Region 3 has the radius approaching twice the mass with time infinitely dilated in the negative direction .In other words time's arrow is reversed as in the tachyon .Region 4 has increasing time with regard to past and future time cones .The isotropic coordinates of Schwarzchild Space-time Metric is from the line element ds2=-(1-M/2r)2 divided by 1+m,/2r)2   and dt2+(1+m/2r)4{dr2+r2)d(omega)2 where omega is the Hubble Constant .Using the Kuskel Extension of Schwarzchild Space-time one has a symmetrical hourglass or cone shaped(spiral)configuration with the throat of the hyper surface at t=0 or infinitely dilated with the radius=twice the mass which is actually a 2-sphere or two dimensional hypersurface with one dimension suppressed. The topologic configuration of the hypersurface is RxS2 and is a circle shown with the radius equaling twice the mass. The surface above the throat at radius=2M lies in Region 1 and below the throat with r=2M in Region 4.There is a hyperbolic region proximal to the throat at the event horizon where r approaches 0 and time approaches infinite dilation.It has previously been determined by this author and the behavior of tachyons that negative mass goes backwards in time and reverses time's' arrow.Region 3 must be composed of negative mass as time's arrow is reversed as time approaches infinite dilationto the negative side while with ordinary matter(positive mass)it approaches time to positive infinity. SUBSTITIUTING a – MASS for positive mass in region 3 and reflecting it back to region 2 of Schwarzchild Space-time the t=infinity and t=-infinity cancel to time=0(zero)and the r=2mand r=-2m(minus 2 Mass)at zero space-time at time=0(zero),it can be concluded that mass and its energy equivalent cancel in regions 2 and 3 and only regions 1 and 4 are left.The information is not lost or destroyed but cancelled out with the negative mass cancelling the positive mass at the event horizon.The quasar effect is scattered throughout space from the 2 dimensional hypersurface over 2(pi)radians or 360 degrees of arc from a point in spacetime of zero entropy and dilated time over string sized space-time into curved Lorenzian Space-time with entropy being increased according to the second Law of Thermodynamics. The above is an explanation for "The Hawking Paradox"which states that the information absorbed by a black hole is lost(energy and mass)when the black hole eventually evaporates.

The negative mass must also produce antigravity and reciprocal curvature of space-time in region 3 which when superimposed on region 2 will push region 3 into region 2 in a form equal but opposite to region 2 resulting in asymptotic flattnessin regions 1 and 4. If one substitutes –mass for +mass in ds2(1-m/2r)2divided by (1+m/2r)2dt2 gets ds2=(1-m/2r)2divided by (1+m/2r)2times (1+m/2r)2divided by (1-m/2r)2 squared dt2.Taking the square root of both sides with negative mass substituting for positive mass with the Schwarzchild metric one gets ds2=1(dt)2or ds2=dt2 as the dx2+dy2=dz2 of the line element cancel out with the negative and positive masses at the event horizon with the negative and positive exteme masses at the event horizon from region 2 and region 3 being reflected upon region 2. This is consistent with space-time approaching zero at the event horizon and stops time at a point of Planck Length where the positive mass from region2 and negative mass of region 3 meet. Note in the cone approaching region 2 and region 3 from the opposite direction the radius gradually approaches zero and the

Rked on the infinite dimension Hamilton-Jacobi Equations with regard to   29
viscosity solutions in an article 11.

mass and negative mass are sequestered proximal to the event horizon so the extreme progressive curvature of space-time to the infinite curvature of a string sized point is preserved. Therefore the net information going in and out of a black hole event horizon is a symmetric bi conar surface with region 3 cancelling region 2 leaving regions 1 and 4 to produce the quasar effect.(diagram eclosed)

CHAPTER SIX:

WHAT IS THE 'EQUATION OF EVERYTHING'?

There exists a simple mathematical relationship between space time and mass relating to gravity(space-time curvature metric) and the speed of light. In terms of a verbal description the relationship is simple. In terms of Relativity and Quantum Mechnics, the relationship is more complex.
As an object approaches an area of extreme mass such as a black hole, time slows down and eventually stops. As any object with mass approaches any other mass with is larger,time slows down even infanitesimilly. As a space-time with an atomic clock would approach the star(Solaris)or the sun, an internal clock would lose at least 1/10th of a second ,possibly more. And if possible to approach a black hole ,time would slow down through dilation towards zero where time would stop .Also space-time=space-time therefore space-time=space-space-time curvature metric(known as gravity) with inertial mass expressing the curvature. Therefore space-time is directly proportional to space. Also space-time is inversely proportional to mass as is proven by the action of space-time as it approaches the event horizon of a black hole. Therefore, space-time is directly proportional to space and inversely proportional to mass. The actual equation would be space-time=space/mass times a constant. This constant is 1/c2 or 1/the speed of light squared as energy=mc2 and the denominator having mc2 becomes the Grand Unification Energy at the point of the "Big Bang" incorporating everything.
In terms of tensors R a b c d=R a b c-1/2 R gab/R a b where R a b c d is curved space-time or Lorenzian Space-time R a b c is flat Mintkowski or Riemannian Space-time R g ab is the space-time curvature metric known as gravity for the metric g ab whose inertial mass is described by R a b which is the Ricci Tensor of that mass.That ENTIRE EXPRESSION IS MULTIPLIED BY THE CONSTANT   1/c2 to give the equation of everything.

The Equation of Everything in terms of Relativity as postulated by Albert Einstein is that all motion is relative and not absolute. When mass(m)travels at approaching the speed of light boundary inertial mass approaches infinity ,space-time approaches zero(0) and length shortens to infinitely short or perhaps Planck Length according to the Lorenzian Transformations. As item A of mass m travels west in an environment that is traveling at velocity B when item A is traveling at velocity A the total velocity is the sum of A plus B.with vectors equaling the components of the motion away from 2 pi radians or 180 degrees such as (A plus B)cosine theta where theta is the angle in radians which is the difference between 2(pi)radians and the net angle displacement of A and B with regard to the surface or manifold .If there is another manifold or surface which is traveling at velocity C which if positive is added to velocities A and B cosine theta If C is negative or traveling less than velocity(magnitude and direction)A and B then velocity C is subtracted from velocities A and B cosine theta. If the manifold is moving in an expansion with a trajectory of 180 degrees as in the "post Big Bang" cos 180 degrees is one so the result would be velocity A and velocity B cos theta +or – the velocity of space-time with regard to the stationary observer. The Equation of Everything is spacetime=space/mass times 1/c2 or $\mathbb{R}$a b c d+R a b c-1/2R g a b/R a b all times 1/c2. In terms of the metric g ab Lorenzian Curved Space-time or Riemann space-time is $\mathbb{R}$ a b c d as previously mentioned .R a b c describes flat space-time on which the metric of gravity R g ab curves space into curved space-time(as previously mentioned).The metric of gravity emanated from mass m whose inertia is described by R ab(Ricci Tensor)and this curves flat Mintkowski Space-time either inward or outward depending on the mass being acted upon by the metric of the mass doing the curving. The sum is described as R g ab where g a b is the metric of mass m.

The apace-time curvature metric of Einstein emanates from Einstein's Equation of Relativistic Gravity where a progressively increasing mass as described by the Ricci Tensor R a b-1/2 the gravity or space-time curvature metric which also increases with increasing mass as space-time curvature approaches infinity as in a point as in the space-time of the pre Big Bang quantum bubble(if below Planck Length)then a quantum foam described by zero-Branes (as previously mentioned).This depends on whether de Sitter Space has a hard boundary at Planck Length as in String Theory's Oribifold. The orbifold again is a twisted cone which is complex and can twist into a CalabiYau Manifold or surface which is a continuous surfacein a puckered appearance of a double torus that communicates with other Calabi Yau Manifolds 5 all six dimensional and in motion.

As increasing inertia and increasing gravity do not increase at the same rate as the speed of light boundary is approached space-time progressively curves to a point at v=c with infinite curvature. According to the Lorenzian Transformations infinite mass in dilated timeand reducing space-time with progressive increase in curvature causes inertia to increase at a greater rate than gravity because the metric g ab is acting on progressively increasing space-time curvature which is R g ab and the space-time curvature metric is half the inertia because the tensors are anti symmetric , abelian and space-time follows Bianchi's Identity .The covariant and contra-variant tensors of space-time are abelian with regard to the space-time curvature metric and as mentioned before are anti-symmetric as inertia approaches infinity at velocity approaches "c" without reaching it which is when the Einstein Tensor G ab=0 which is the stress energy tensor T ab at infinite space-time curvature. Anti-symmetric tensors cancel out in magnitude but with opposite direction.
THE EQUATION OF EVERYTHING IN TERMS OF QUANTUM MECHANICS HAS space-time=space/massx1/c2 described in terms of Planck Mass(the smallest unit mass for a quantum particle)or that two quanta can occupy and is approximately

Rked on the infinite dimension Hamilton-Jacobi Equations with regard to 31
viscosity solutions in an article 11.

1.22x10-24 kg .Mathematically Planck mass is the square root of hc/8(pi)G where h=Planck's constant at 6.63x10-34 and G is the Gravitational Constant of 6.67x10-11 newton-meters/sec2 and c=3x10 8 meters/sec and is the speed of light boundary. Space-time is described as the n-Dimensional state of a point-particle x at time t as a probability function and is operated on by the Hamiltonian Operator defined as – h/2mtimes the La Placean Operator with respect to the second derivative or d2/dx2+d/dy2+d2/dz2/ So utilizing the above space-time of H a}(r,t)|2 this is the probability density of a point particle "r" with respect to time or "t" in n-Dimensional Space. This equals -1/2e to the + or – I to the n cotangent theta power,here i=the square root of -1or -1e is the inverse or reciprocal of the natural log which is the integral of du/u which relates to the spiral fractal formula introduced in this authors first book "Mega physics ,A New Look at the Universe" and this defines the ground state in a Relativistic Universe as the natural log (l n 1)=0 and the natural log of infinity=infinity. Theta is the angle of trajectory at Planck Time from "The Big Bang" which is pi radians or 180 degrees. The infinity power of e(2.71828)is infinity and e to the negative infinity power is zero. Therefore e –in cotangent theta power defines the ground state as ln 1=0 and e I n cot theta is a reciprocal function and describes 1/0 which is infinity but e –in cot theta is zero where n=number of dimensions and theta is the angle of trajectory or pi radians.Therefore Planck's Mass(c2)H a|(r, t)|2 d n r where this expression is multiplied by the La Placean Operator in the n-dimensional state. A is the number of eigenstates of energy operated on by the Hamiltonian Operator (-h 2/2m)x La Placean Operator.In this case the Operator defines the wave function of r with respect to time(t) and also explains weak perturbations which explain quantum fluctuations in deSitter Space .Note also that the Hamiltonian Operator-minus h2/2mtimes the LaPlacean Operator)2+V o or initial velocity reflects momentum p=mv where p(rho) is Momenetum. So the momentum of quanta with Planck's Mass is incorporated over n-diemsnions from a to n eigenstates of energy and weak perturbations must include the momenta of quanta from the a=0 to a=n eigenstates of energy or energy levels. Therefore c2(hc/8 pi G)1/2+H a(eigenstates)|(r,t|)2 d n r times the LaPlacean Operator represented by the inverted delta to the n power equals =-1/2e +or –i n cotangent theta power with fluctuations which is again described by the Hamiltonian Operator in "a"eigenstates in the n dimensional state and ei n cot theta power is infinity. Therefore as 10 19 GEV approaches infinity.here n is the number of dimensions and I is the square root of -1. The Hamilton Operator in n dimensions describes the wave function of a point particle r with respect to time(t) in the n dimensional state with respect to "a" eigenstates of energy.c2(hc/8 pi G)1/2=10 19 power Giga Electron Volts which is the Grand Unification Energy which includes all forces except weak perturbations from quantum fluctuations with are corrected for by the Hamiltonian Operator. This occurs in the multiverse where the n-dimensional state approaches infinity and can be proven by subdividing a sphere to one second of arc or 1/3600 of a degree. This second of arc can be subdivided down to infinitiy and each subdivided portion is in motion as part of space-time with an infinite number of intersections of each unit or infinitely subdivided second of arc. As the intersection of two planes define a dimension and since the subdivided portions are not parallel due to space-time curvature when any

Rked on the infinite dimension Hamilton-Jacobi Equations with regard to 32
viscosity solutions in an article 11.

even an infinitesimal amount of mass(non-vacuum)curves all space-time there are an infinite number of intersections from these subdivided lines or planes indicating an infinite number of dimensions in the multiverse.

When one second of arc is a plane in motion the topological surfaces mimic an osculating plane 7and each osculating plane from a topological standpoint define a dimension as space-time curvature causes an infinite number of intersections of these infinite osculating planes. According to Zeno's Paradox each degree is subdivided ad infinitum that prevent two solid objects from touching. With an infinite number of dimensions for space-time the left side of the quantum mechanics equation narrows from 10 19 giga electron volts toward infinity without ever reaching it as is true with the right side of the equation. There are also other theories such as M Theory which states the zero-branes which were building blocks to everything may have incorporated an infinite or near infinite number of dimensions .This situation applies for space-time to -1/2e –in cot theta and this applies to everything with it's reciprocal -1/2e i n ot cot theta power and this applies to zero space-time on the right for -1/2 e –I n cot theta. On the left side the Planck Mass is zero in the ground state as this is the vacuum or null state so c2(hc/G)1/2=0 and the zero-dimensional state states that the Hamiltonian Operator or point particle|(r,t)|2 d n r times the La Placean Operator in the zero dimensional state is also zero because the derivative of zero is zero. Therefore 0=0 in the ground state and the Equation of the Universe or Multiverse is upheld with respect to Quantum Mechanics. Note that this expression equals the Relativity Expression ℝ a b c d=R a b c-1/2R g a b/R a b times 1/c2 and due to the transitivity postulate that a=b,b=c therefore a=c indicates that the Quantum Mechanics Equation which is zero in the ground state and the Relativity statement which is zero in the ground state(as R a b c d=0 as Riemannian space nets zero when mass or inertial mass R a b approaches zero. Of course R a b c-1/2R g ab=0 ; R a b approaches zero forms the expression 0/0 which is everything as the case of curved Lorenzian Space-time. This is the case of where approaching the 0 dimensional case incorporates everything.

Kurt Godel described a scientific tenet called "The Axiom of Incompleteness "stating that in any axiomatic system the set containing all elements must be incomplete. If there exists a set containing this subset ,the axiomatic system must be incomplete. Based on this tenet "An Equation of Everything "must be incomplete although it appears complete. An example of this is when 10 19 Gigaelectron volts(The Grand Unification Energy of all energies from "The Big Bang")can only approximate the value of infinity and while it is true the 10 19 Gev approximates infinity due to weak perturbations from quantum fluctuations it isn't definitive. However in the infinite dimensional case where 0=0 as 10 28 power in the denominator of the left side and infinity in the denominator of the right side approximate 0=0.This will again be brought into focus later. Quantum Mechanics and Relativity are unified with the equations ℝ a b c d=R a b c-1/2R g ab/R ab times 1/c2 and Planck Mass times the expectation value of the probability of a point particle r at time t in the n dimensional state where space-time is -1/2 e –i n cot theta with the reciprocal curvature being -1/2e i n cotangent theta where n=the

Rked on the infinite dimension Hamilton-Jacobi Equations with regard to viscosity solutions in an article 11.

number of dimensions netting infinity divided by infinity which is everything except zero which is the null state or absolute vacuum state indicating in this total case that nothing or spaceless ness doesn't exist. Based on measurements the integral of du/u mathematically indicates -1/2 e –i n cotangent theta power as space-time with the curvature metric R g ab while the sum total of both reciprocals for space-time would result in flat space-time as in a vacuum which is precluded by the expression infinity/infinity because it doesn't include zero and is therefore incomplete .The Quantum Mechanics equation of everything

is $\dfrac{\dfrac{\psi(r,t)dn(power)r\nabla n}{c2\left(\frac{\hbar c}{8\pi G}\right)1}}{2} + \mathcal{H}\ from\ a\ to\ n\ eigenstates(|r,t)|\ d\ n(power)r\nabla n =$

$\psi(r,t)d\ (n\ power)r\nabla(n\ power) - \frac{1}{2}e - i\ n\cot\theta$ equals the Relativity Equation R a b c d=R a b c-1/2R g a b/R a b times 1/c2 and at the ground state they both equal zero and each other and can be simplified to space-time=space/mass all times k=1/c2 where c=3x10 8 meters/sec.

The Hamiltonian operator

$\mathcal{H} =$

$\hbar\dfrac{2}{2m}\nabla\ to\ the\ nth\ power\ where\nabla\ (nabla)is\ the\ LaPlacean\ Operator$ handle the weak perturbation

or quantum fluctuations acted upon by the wave function of the point particle r at time t. This will be delved into again later in this book.

Another equation postulated was the wave function$\psi(r,t) = \int e^{\frac{i}{\hbar}}$ to the integral

power$\int\ (\dfrac{R}{16\pi G} + \dfrac{1}{4F2} + \psi iD\psi - \lambda\varphi\psi\psi + D|\varphi|2 -$

$V(\varphi)where\ \psi is\ bar\ \psi\ or\ a\ probability\ function.This$ is based on the Yukana Equation, Relativit

The Schrodinger Equation and the book "Quantum Mechanics and Path Integrals by Richard P. Feynman and Albert R. Hibbs. The path

integral

$\oint\ \nabla 2\psi(r,t)applies\ Poisson'sEquation\ for\ the\ path\ of\ dual\ vector\ field\ 4\pi\rho\ where\ \rho\ is\ the\ en$

Energy density of matter and

$\psi(r,t)is\ the\ wave\ function\ of\ point\ particle\ r\ at\ time\ t forming\ a\ path\ integral of\ 16\pi\rho 2\psi(r,t)$

$\dfrac{16\pi\left\{\frac{\rho 3}{3}\right\} -}{3\psi(r,t)superimposed}\rho 3$ with parallel transport to curved manifold (surface)with space $-$

$time\ curvature\ of\ -\frac{1}{2}e - i\ n\ cotangent\theta\ for\ \frac{\rho 3}{3}\ and\ space -$

$time\ curvature\ of\ -\frac{1}{2}e +$

$i\ n\ cotangent\ theta\ for\ -\frac{\rho 3}{3}\ again\ approaching\ the\ ground\ state.The - |D\phi|2 -$

Rked on the infinite dimension Hamilton-Jacobi Equations with regard to 34
viscosity solutions in an article 11.

CHAPTER SEVEN
WHAT IS M THEORY?

M Theory is a combination of the five extant string theories ;type I, Type II ,Type IIa
,Heterotic 8x8 and the SO(32) string theories. The five string theories are dual to
each other as previously mentioned mathematically observing the same
phenomenon from five different vantage points or approaches all describing the
same thing. What makes M Theory the combined five string theories is the
incorporation of membranes which vibrate. These membranes are submicroscopic
and may be string sized 10-33 cm or possibly smaller. A membrane is almost
continuous with any and all matter and possibly energy and are continuous across
the different dimensions whether 26 compactified(curled up)to 10 as in type I string
theory or an approaching infinite number of dimensions if the osculating plane can
be applied to an infinite number of non-parallel planes which are subdivided from 1
second of arc(1/3600 degree)and in motion expansion and rotation as in Godel's
Rotating Universe .These non-parallel planes must intersect an infinite number of
times if they are non-parallel due to the space-time curvature metric forming an
infinite number of dimensions in the osculating planes. Membranes would have to
be contiguous with these osculating dimensional planes and would contain all
matter including energy which is converted to matter as matter=energy/c2. M
Theory was originally coined for Membrane Theory and Matrix Theory depending
on the source read and the different membranes called in short hand branes
describe different states of matter interacting with energy in space-time .M Theory
has no clear cut definition except the duality with all five string theories although
when utilizing super gravity for the existence of the 11th dimension instead of the
compactified 10 dimensions in the low energy limit involving D-0-Branes it reduces
to type II a string theory when compactified by
2πR where R measures the limit to infinity of D − 0 −
Branes in the infinite momentum limit of M Theory as in the case of U(N)in the super Yang Mi

Theory 9. The compactified type IIa string theory is a sphere of Radius R. The D-0-
Brane has a momentum limit of 1/R and a bound state of D-0_branes have a
conditional momentum of N/R where N relates to the unified Yang Mills state
relating to the gauge limit in terms of d4space such that g2YM N approaches infinity.
Note that D-1-Branes or 1-branes describe the string as well as the monopole
depending on the text read.M theory incorporates super gravity and the 11th
dimension to string theory."branes "'are movable membranes that incorporate
special dimensions and particles including strings. Membranes as mentioned
previously vibrate move and are real and measurable .Charges ,state and tension
are incorporated into membranes which extend up to at least the ten dimensions of

Rked on the infinite dimension Hamilton-Jacobi Equations with regard to   35
viscosity solutions in an article 11.

string theory. The limited definition of M Theory is "the limit of strongly coupled IIA string theory with 11 diemsnional supergravity or the Poincare invariance"The 2-brane or M 2-brane couples to the potential(V) of eleven dimensional supergravity .The 5-brane or M-5 brane carries an electromagnetic charge of the potenetial(V)of eleven dimensional supergravity.The 5-brane carries the potential which is coupled to the 2-brane. The Neveu-Schwarz 5-brane carries the electromagnetic charge(V)of what is known as the NS-NS 2-form potenetials where the NS boundary state is a fermionic field that is anti-periodic on the world sheet in the closed string or the double of the open string. The Neven-Schwarz algebra is a world sheet algebra with regard to the energy momentum and super-current tensor into a sector where supercurrent is antiperiodic and modes are half integer valued .Therse are what are known as Fourier modes .The null state is orthogonal to all physical states including its' own. Incorporating the anti-periodic fermionic field which is the NS boundary state onto the 4 and 5-brane representing space-time in the four macroscopic dimensions would have the Neven-Schwarz 5 brane incorporated with the NS-NS 2-form potentials on the world sheet(which represents flat or two dimensional matter) and is divided by the Ricci Tensor with regard to the sum total or infinite sum of all inertial mass to equal the Lorenzian tensor of space-time in terms of supergravity incorporating 11 dimensions. The momentum limit of the inertial mass as it approaches the infinite momentum limit where "R" approaches a very large number is incorporated into the denominator of space-time=space/mass .Space-time curvature based on the infinite momentum limit R where N approaches infinity in the unified Yang Mills state in the d4 state relates to the macroscopic gauge limit and the D-0-brane which is bound with conditional momentum of N/R. The momentum limit of 1/R of the D-0-branes relates to the infinite momentum limit and curved Lorenzian space-time or Riemannian space-time where the 256 permutations net out to zero in the ground state. Applying this to 11 or possibly more dimensions(infinite dimension state as previously mentioned in terms of M theory is more difficult.

SUPERGRAVITY which incorporates the 11th dimension into superstring or the IIa closed string theory to compose M Theory where the IIa closed string theory compactifies to a circle or sphere is involving global supersymmetry which produces a Yang Mills Gauge Group of Osp(1/4).The number of states within 11 dimensional supergravity are eAM implies ½(9)x(10)-1=44 ψM *implies* $\dfrac{1}{2(9x32-32)} =$

128 *and A MNP* $\begin{pmatrix} 9 \\ 3 \end{pmatrix}$ =84.where M and N represent 11 dimensional curved space indicies with 32 dimensional spinors. The term e A M where A is the superscript and M is the subscript and it represents a linking or parallel transport of the base manifold or surface with the tangent
space.

ψM is the graviton field and A MNP is an antisymmetric tensor field with 128 *boson fields eq*

equalling the fermionic field giving a total number of 256 boson and fermionic fields which is the number of fields in the 11 dimensional N=8 model for super gravity again where bosons and fermions curve space-time.

Rked on the infinite dimension Hamilton-Jacobi Equations with regard to 36
viscosity solutions in an article 11.

As there are 256 bosonic and fermionic fields curving flat Mintkowski or Riemannian space-time which has 256 permutations and since Riemannian space-time graviton or bosonic fields are anti-symmetric tensor fields acting on flat space-time as anti symmetric tensor fields the net permutations of gravity space-time curvature on flat space net out to zero which is the numerator of space/mass=space-time .Regardless of the denominator even when multiplied by the speed of light squared the net result is zero which is curved Lorenzian or Riemannian Space-time.Space-time in the "null state "or equivalent to the Einstein Tensor G ab. Therefore "the Equation of Everything" does apply to 11 dimensions as incorporating supergravity. Note in the above Riemann Space-time can be applied as curved or flat when Mintkowski Space-time is always flat.Riemann postulated that all mass can be described as curves in space-time although flat space-time means that a missal shot out in flat space will theoretically never return to the starting point but with curved space-time it will eventually return to the starting point as in Einstein's Closed Curved Universe vs. Friedmann type II open flat expanding universe.

In terms of M-Theory the Neven-Schwarz 5- brane incorporated with the NS-NS 2 potentials incorporating the anti-periodic fermionic field(vacuum state of ferminon field) on the world sheet would go into the numerator of space/mass and the infinite momentum limit in the bound state of D-0-branes would go into the denominator suggesting the inertial mass times the speed of light squared of the quantum bubble with approximately 750 billion strings acting on D-0-branes in the pre Planck Time epoch of under 10-43 seconds. The world sheet would be 2 dimensional as are strings and the anti-periodic fermionic field density in terms of a tensor field and applying Poisson's Equation for dual vector fields   would be applied via parallel transport onto the flat manifold of the world sheet. Since the energy momentum and supercurrent tensor are also anti-periodic the energy momentum R goes into the denominator with the supercurrent incorporating into the 1-brane as a carrier such as the monopole or electron cloud while the fermionic field density is incorporated in the numerator as previously mentioned. This incorporates to enclose a near infinite momentum onto an accelerating expanding space-time acted upon by gravity (fermionic fields caused by the inertial mass of the quantum bubble).AS A RESULT N-BRANES SUGGESTIVE OF LORENZIAN SPACE-TIME=D-0-BRANES+ and –Neven -Schwarz 5- Brane incorporated with the NS-NS 2 potentials/N/R where R is the infinite momentum limit and N=1. The N-Brane as an infinite number of dimensions are approached but never reached approaches infinity but is never reached and since
anything/1/
∞ is infinity due to the infinite momentum limit so as applied to N dimensions or the N − brane infinity =
infinity with supergravity localizing the Einstein Tensor G ab which is $8\pi T$ which is the stress

Energy tensor of matter acting by the Poisson Equation and is zero(0)based on the D-0-branes.

Rked on the infinite dimension Hamilton-Jacobi Equations with regard to viscosity solutions in an article 11.

CHAPTER EIGHT
WHY THERE IS AN ARGUMENT THAT THERE ARE AN INFINITE NUMBER OF
DIMENSIONS

   A dimension is formed by the intersection of two planes which have an
intersection of an angle which in length, width and height is 90 degrees or pi/2
radians. However other dimensions may have different angles of intersection which
aren't perpendicular .For example ,the intersection of the x ,y ,and z axis has a near
infinite number of points which while discontinuous can be subdivided down to an
infinite number of intersections and the limit of these subdivisions can blend or blur
into a continuous dimension. There are a postulated 26 dimensions associated with
string theory and superstring theory and these dimensions can be
compactified(rolled or curled up)to 10 dimensions or 11 dimensions with
supergravity included. To make a corollary in logic one must depend on axioms as
assumptions. These axioms must be assumed to be true and if false the corollary can
be false. For example if it's an axiom that planes are stationary and later it is found
that planes are in motion, then the corollary that there are three dimensions length,
width and height with no others as in Euclidian Geometry is false. In the case of
space-time this corollary is false as time is considered a fourth dimension by Albert
Einstein and his space-time curvature theory which was proven with the "solar
eclipse" experiment in 1921 to measure a change in position of an object near the
sun(which is of significant mass compared to empty space)proves that space-time is
curved and not flat(without curvature).It was mentioned previously by this author
that each subdivision of a second of arc(or $1/3600^{th}$ of a degree)can be subdivided
down toward an infinite number of cuts. If these cuts of space-time were totally
parallel to each other as if space-time were totally flat and without curvature, then
these cuts would never intersect each other. However because space-time has been
postulated as being in motion along with mass, expanding and rotating(according to
this author and Kurt Godel),then these subdivided planes would not be absolutely
parallel but would be almost parallel, where the parallel nature is distorted as space
approaches a reads of extreme mass such as a black hole where the curvature of
space-time increases towards infinity which is why space-time appears to spiral into
a black hole in a cone like configuration as the event horizon is approached. The
time component or fourth dimensions shrinks toward zero as the extreme mass is
approached as space-time is inversely proportional to mass.
   Now if there are a near infinite number of planes from the subdivisions of each
second of arc and is these planes are osculating and in motion expanding with a
progressive decrease in rotation from the area of maximum rotation at just before
Planck Time 10-43 seconds and approaches pure expansion with only a miniscule
rotation as 13.7 billion   years later ;then these planes are osculating into expansion
and rotation vectors with the unit tangent vectors being measured along these
osculating planes. Obviously ,if all these axioms are true ,these near infinite(if there
is no downward boundary to space-time at below Planck Length or 10-33 cm)and if
the quantum foam can also be subdivided down to infinity then there would be an

Rked on the infinite dimension Hamilton-Jacobi Equations with regard to   38
viscosity solutions in an article 11.

almost infinite number of intersections between the near parallel planes formed by each subdivision from 1 second of arc downward toward infinity .As Einstein showed that these planes of space-time are curved and "The Big Bang" Theory shows them in motion ,these are osculating planes and the unit tangent vectors reflecting the curvature of these planes and the planes will intersect a near infinite number of times forming a near infinite number of dimensions. This was all explained earlier in this book and there are other arguments to this premise as well. This conclusion would be totally consistent with the infinite or near infinite universe theory and would preclude that D-0-branes are not at absolute zero dimensions but an infinitely small number approaching the zero dimensional state as it is when an object approaches 0 degrees kelvin but cannot be reached .Indeed the null state or D-0-branes are only an approximation where the zero dimensional state is approached as an asymptotic function as with limits where the size of each extant dimension is reduced toward zero without reaching zero. The reason why the axiom nothing doesn't exist when according to "The Big Bang" ex-nihilo states it does is because of these limits in size downward for the dimensions which are in motion with the osculating planes never reaching the zero dimensional state as when "time's arrow" reverses time from before the Big Bang when a Big Crunch would cause time to move backwards in a massive implosion bringing it back toward zero time without ever reaching it before the Big Bang moves time's arrow forward again. These subdivisions can be curled up or compactified toward zero BUT ARE NOT ZERO ONLY APPROACHING ZERO AS AN APPROXIMATION .Also the diverse motions of string and superstrings in space-time are diverse but the number of motions from a vibration to a twist,partial rotation(degrees can also be subdivided plus the motions of strings can be subdivided)make the two dimensional world sheet as an approximation as would the two dimensionality of flat matter or flat anti-matter .Indeed, the other dimensions are there but so miniscle and compactified that for the sake of math they aren't important and there limits can be brought to zero.
Infinite Dimension Symmetry involves the symmetry transformation of space-time as. generated by similarity transformations of T ab which is the stress energy tensor of theories associated with conformal gravity C.A type of algebra exists called infinite dimension subalgebra which is associated with Lie Groups as a subgroup of Lie Algebra. Infinite dimension subalgebra has nonsingular commutators and uses gauge symmetry of the Yang Mills groups to form a weighted tensor algebra. An approaching infinite set of conformal fields have well definied commutators of an arbitrary pair of zero modes constituting an infinite dimension subalgebra with regard to the full symmetry of string theory and M Theory(including the compactified IIa string theory which is spherical .Infinite dimensions do not arise through the moding of a finite number of conformal fields but rather than an infinite number of conformal fields of space-time acted upon by any
mass.W
∞[17]*in terms of field equations with Wi*[18]*as conformal weight W∞ must retain all modes (*
time space −
time while zero modes form only Cartesisan Subalgebra. This subalgebra possesses the proper

Rked on the infinite dimension Hamilton-Jacobi Equations with regard to   39
viscosity solutions in an article 11.

Ties of string symmetries and is a supersymmtery in that it's generators do not commute with the generators of the Lorenzian Transformations and comingled or quantum groups which are excited and with a different spin it is spontaneously broken in flat space-time because NOT ALL GENERATORS COMMUTE WITH THE STRESS ENERGY TENSOR(T ab)relating to the free scalar and it transforms excitation of differing masses into one another .Symmteries or gauge symmetries should be local symmetries and the constant tensors
ψ and χ in the generator were depending on the scalar field $X\mu(\sigma)$ where mu is a superscript th

Locality would be apparent,nut it is x dependence that affected commutations. Infinite dimension subalgebra would be a global part of Gauge Symmetry and must include both holomorphic and anti-holomorphic derivatives with propagating degrees of freedom generated by operators evenly balanced between two types of derivatives.Short distanc e singularities e ipxx L(x)ei qXL(w)equals the fraction e i(P+Q).xL(W) /(Z-w) to the-p .q power and e ip.x(=L(x) is also a power with e iqXl(w) as a power.The stress energy relates to (Z-w)to the –p.q acting upom e i(P+Q).Xl(W) which relates to e –in cotangent theta of space-time, The infinite dimension state where Napproches infinity forces the N-brane of M theory to approaches
an ∞ Brane for Curved Lorenzian Space $-$ time R a b c d and the superequation

{Ha to n eigenstates|(x,t)2d nr
∇n divided by $c2\sqrt{}\ \hbar \frac{c}{8\pi G} =$
$\nabla n|x,t|2d\ n\ x\nabla x\left(-\frac{1}{2e}-\right.$
in cotan theta$\Big)$ where n dimensions approaches ∞ therefore $-\frac{1}{2}e-$
$i\ n\ cot\theta$ approaches $e - \infty$power so the wave $\dfrac{function\psi|x,t|2d^{\frac{n(power\ of\ )x\nabla n}{-1}}}{2}e-$
$i\ n\ cot\ thea = \psi|x,t|2dn\frac{x\nabla 2}{\infty} = 0.\ Here\ space - time\ is\ described\ as\ -\frac{1}{2}e-$
$i\ n\ cot\theta$ where theta approaches π radians or 180 degree trajectory of a sphere describing

the Big Bang. Note the particular math of infinite dimension su-algebra gets extremely involved and can be referred to by the footnote 10.

Call{ } the set of all covariant tensors in d –dimensional space.The operatorΔr acts on the subset$\{\psi(k),w\ i\}$ such that $\Sigma l = 1$ to $r\{\psi(h),w\ i + \delta il\}$ where ψ is the wave function of w i and here i is the initial event the element of algebra of

Rked on the infinite dimension Hamilton-Jacobi Equations with regard to viscosity solutions in an article 11.

Pairs such that$\{\psi(h), w\,i\}$ *modulate relations* $\rightarrow \lambda\{\psi h, wi\} + \omega\{\psi(h), w\,i\} - \{\lambda\psi(h) + \omega\psi(h), w\,i \rightarrow)$ *such that* $\Delta h\{\psi(h), w\,i\} \rightarrow 0. \psi$ *is a subset of S.* In this case h is not Planck's Comstant but a variable power to the wave function. S, η *is the Mintkowski Metric for flat space* $-$ *time.* $|S|$ *are the elements of* $S; S - \psi$ *are the complement of* ψ *M.S. Conformal weights* $\omega 5 - \psi + V(t) - P(u)$ *were conformal weights of* ψ *that don't correspond to the indicies in* ψ *together with the* ɩ

Weights of
χ *that don't correspond to the indicies in* ψ *with the conformal weight of* χ *that don't correspon*

To indicies in the image of the wave function of w and P(U) such that $\omega s - \mu + V T =$
$P(U)$ *Wick's Theorem states the sum over* μ *and P is a sum of all possible contractions of the te*

Tensors
ψ *and* χ *The conformal weights are those of the noncontracted indicies except the weight*

Of
ψ *and are increased incrementally by the weights of the contracted indicies through the powe*

Of the operator
$\Delta | S -$
$\mu |$ *The two point function for free bosons implying the curvature of Mintkowski space* $-$ *time by any mass is* $< x\,m(z)xV(w) \geq$
$-\log(Z - w)\,\delta vto\mu$ *where* δ *is the superscript and* μ *is the subscript and the* $-$
$\log(z -$
$w)$ *is to this* δ *where* μ *is the covariant tensor and* δ *is the contravariant tensor all aplied to the* ɩ

Boltzman Equation incorporating states of matter with regard to bosons and conformal gravity. Wick's Theorem states that short distance singularities as at the event horizon of a black hole form a conformal block or non-holomorphic operator or sum of the products of holomorphic and anti-holomorphic fields constructing a full commutator out of the commutators for anti-holomorphic components .Holomorphic and anti-holomorphic operators are constructed from mutually

Rked on the infinite dimension Hamilton-Jacobi Equations with regard to viscosity solutions in an article 11.

commuting sets of the creation and annihilation operators .Short distance
singularities at Black Hole Event Horizon are eip.xL(z) where ip and x L(Z) are
powers times e iqX L(w)=the quotient of e to the i(p+q).xL(w) power/(Z-w)-p.q .p.q
must be integers .Products of holomorphic and anti-holomorphic fields constructing
a full commutator out of commutators for anti-holomorphic components form the
conformal block on space -time|S| giving a mechanism in terms of math on how
space-time is crammed down by annihilation operator in terms of bosonic effect of
extreme gravity on asymptotically flat Mintkowski Space-time
s,η as event horizon is a black is reached. This may appear far a field from infinite
dimensions but the stress energy tensors relating to the annihilation and creation
operators acting on Mintkowski Space-time clearly reveal that space-time or the n-
brane =space(D-0-Branes + or – the Neven Schwarz 5-brane with NS-NS2
potentials/N?R where R is the Infinite Momentum potential as N approaches
1(which is mass or the Ricci tensor x c2).They also explain why space-time spirals
down to a point (string sized?)without actually disappearing. The Hamilton-Jacobi
Equation for infinite dimensions is such that $\lambda v(x)+< Ax = \phi(x), Dv(x) >$
$+H\big(A\mathcal{B}power\ x, D\ v(x)\big) = 0$ where $x \in X$ wher e X is a real Hilbert Space, $\lambda >$
0 and $H\colon H$ by X is continuous as a closed linear operation with a compact and dense inclusion I

D(A) is less than X. We assume A is positive and self
adjoint.$\phi\colon D(A\ \beta\ power)implies\ D(A\ -$
beta power)and is Lipschitz continuous. Piermarco Cannarsa and Maria Elisabetta Tessitore w

CHAPTER EIGHT

The infinite dimension Hamilton-Jacobi Equation and applied specific situations is in

a paper written by Piermarco Cannarsa and Maria Elisabetta Tessitore called

"Infinite Dimensional Hamilton-Jacobi Equations and Dirichlet Boundary Control

Problems of Parabolic Type".

The stress energy tensor of a free scalar T a b$\longrightarrow T(\sigma)$ and $T(\sigma') = \frac{1}{2} : \partial x \partial x(\sigma) :=$

$$\lim \epsilon \longrightarrow \frac{0i}{2\partial x(\sigma)\partial x(\sigma+\epsilon)} + \frac{1}{4\pi\epsilon 2} \; where \; the \; commutator \; is \; 4[T(\sigma), T(\sigma')] = \lim \; \epsilon, \epsilon' \longrightarrow$$

$0[\partial x(\sigma \partial x(\sigma + \epsilon)]\partial x(\sigma')\partial x(\sigma' + \epsilon')\}. \, for \, real \, numbers. \, or \, F(\sigma\delta(\sigma - \sigma') -$

$f'(\sigma')\delta(\sigma - \sigma') \longrightarrow$

$[T(\sigma), T(\sigma')]which \, applies \, to \, the \, commutation \, of \, the \, stress \, energy \, tensor \, where \, \sigma \, and \, \sigma' have\iota$

E stress   energy as related to T a b where g a b relates to sigma and sigma prime.

Incorporating imaginary numbers $2iT(\sigma')\partial'(\sigma - \sigma') - iT'(\sigma')\delta(\sigma - \sigma' - \epsilon')/(\epsilon +$

$\epsilon')2 + 2\delta(\sigma - \sigma') - \delta(\sigma + \epsilon - \sigma')/(\epsilon + \epsilon')3.$ Stress Energy between   sigma and

sigma prime yields epsilon relating to a small number for stress energy relating to black hole entropy(s) using Virasuro Algebra: e ip .$X(\sigma)$ :: $e\ iq.\ X(\sigma + \epsilon)power =$

: $e\ ip\ dot\ X(\sigma) + iq\ power\ dot\ x(\sigma + \epsilon)$: $e\ p -$

$\frac{q}{2\pi}power$. $As\ \epsilon\ is\ greater\ than\ zero\ the\ entropy\ of\ a\ black\ hole\ approaches\ zero\ by\ the\ normaliz$

normalization coefficient with normal ordered operators .The stress energy

between sigma and sigma prime is a maximum at the event horizon or a black hole

where space-time approaches epsilon.11

As is well known
s=r(theta)
$\theta\ where\ s\ is\ \ the\ \ distance\ or\ arc\ length\ subtended\ by\ a\ sphere\ r\ is\ the\ radius\ of\ the\ cut\ of\ the$

If the sphere is compactified and converted to two dimensions with regard to the

world sheet as in type IIA string theory converts to a description in which M Theory

may be derived .As theta

$\theta\ approaches\ zero\ degree\ but\ not\ reaching\ it\ the\ cuts\ becomeinfinitely\ small.\ The\ radius\ rema$

remains constant. As the sphere describes space-time in the Big Bang rotating and

expanding according the H the Hubble Expansion Coefficient the cuts "s" approach

zero initially then expand as theta approaches zero then expands and "r" describes the radius of this universe .If the near infinite number of cuts remained parallel

theta would remain very small (below one second of arc).The radii would approach

being parallel without reaching it as the angle between radii approach zero degrees

without reaching it. The cuts are not parallel due to the extreme mass of the

quantum bubble with extreme Dark Energy overcoming the extreme mass of the

bubble curving space-time to infinity as in a point which is a two dimensional

sphere composed of strings. The unwinding curvature of space-time will reach the

space-time curvature metric of Einstein instead of that of asymptotic flat space.

These infinitely small cuts with ds approaching zero as theta approaches zero with r

or the radius continuously increasing according to the Hubble Expansion Factor

forms an infinite number of dimensions as the curvature causes an infinite number

of intersections of these non-parallel infinite number of planes formed by the

infinite number of cuts in the circle (2D) or sphere (3D) or space-time (4D) and

these planes are osculating with R and R. moving along the Normalization component such that arc length parameter r=r(s) with s=$\|\int_a^t \left\|\frac{dr}{du}\right\| du$ or $\frac{ds}{dt} =$

$\|r\|$ with a dot over r. r. is differentiation with respect to time r' differentiation with respect to

distance.r.=r'=dr/dt,r..=r"=d2r/dt,r...=r'''=d3r/dt. Mapping t to s has inverse relation

of s to t given by t=psi prime(s) dt/ds =psi prime(s)=1/||r dot||The moving frame is

such that T(unit principle normal)r'=(dx/ds,dy/ds,dz/ds)The Binormal vector with

a curve B=TxN for the cross product shows all plane curves have a principal normal

and these curves are not paralle.N=(-

$\sin\theta, \cos\theta, 0)$ plane $z = 0$ if $T =$

$(\cos\theta, \sin\theta, 0)$ The Moving frame at any point r. isn'tzero but may be episoln and r. r: aren'tzerp

zero then T=r(dot)/||r(dot)||and

N=$\varepsilon(r.r.)(r..) - (r.r..)r./\|r'\|\|\|r.xr..\|$ |where x is the cross product

This moving frame in the form of a triad moves continuous along C where C is
analogous to r is the formula.:
s=r
θ of a sphere divided into an infinite number of moving cuts or planes and each C and N are n

Are mutually orthogonal the triplet of unit vectors T ,N and B constitute a right

handed system of basis elements E3 and cover space-time curvature as -1/2e-in cot

theta .The triad of T,N and B moves continuously along C and is the moving frame or

triad whereby T and N are the osculating plane.(touching unit tangent vectors)12

Now considering compactification of a near infinite number of dimensions or cuts in

a circle or sphere ,the area can form a point with infinite curvature moving at the

rate of the Hubble Expansion factor with the force or energy of the Big Bang.

Therefore the infinite intersections of planes in motion outward with a curvature

metric going from infinity as a point to space-time curvature metric from any mass

as derived from Einstein these dimensions which are the intersection of two moving

planes along the vectors T ,N and B will compactify below Planck Length or 10-33

cm which is the size of a string leaving at least 10 or possibly 11 non-compactified

dimensions as in supergravity or string theory rather than the 26 non-compactified

dimensions .The creation and annihilation operators which are holo- morphic and

anti-holomorphic can be used to show space-time crunching to an infinite curvature

point in a"Big Crunch" from an expanding and rotating sphere and expanded and rotating from a point of infinite curvature to this universe with all infinite

dimensions compressed in the infinite curvature point to where space approaches

nothing without reaching it and the infinite dimensions all MUTUALLY INTERSECT

AT THAT POINT .Again the compactified form of II a string theory becomes a two

dimensional sphere or circle which corresponds to that infinite curvature point or

expands with two dimensional components such as closed strings to THE WORLD

SHEET.

A complex idea being made simple is based on Ockham's (or Occum 's) Razor. All

things being equal ,the simplest explanation tens to be the right one. If a two

dimensional sphere is a circle which is the COMPACTIFIED FORM OF II a String

theory then the equation

$s=r$

θ *where s and θ are subdivided down to infinitely small slices but with a radius that approache*

Infinite length has a compactified infinite number of dimensions of N-branes
approaches the D-infinity Brane where all of the dimensions are compactified in the
quantum bubble the dimensionality approaches the D-0-Brane just as

2

πradians approaches 0 radians on a circle but never reaches it. The radius if approaching infi

Infinite length going across 360 degrees or 2(pi) radians could radiates flat matter (2 dimensional matter or radiation)at v approaches the speed of light boundary where space-time shrinks because the inertial mass of matter approaches infinity .A radiating energy source over 2 pi radians from the quantum bubble as the radius before the "Big Bang" would travel at v approaches c. Length=Length (0)(1-v2/c2)1/2 from the Lorenzian Transformations for matter .There is length contraction at v=c just as there is time dilation where time slows down towards infinity. There is a theory that length was the first dimension de-compactified and would have to be infinite ®R from which the Hubble expansion coefficient would follow but according to the Lorenz Transformations radiation would have to be infinitely short if traveling at v=3x10^8 meters/sec which can happen in collapsed space-time where time is dilated towards infinity and space is compactified to the size of a quantum bubble or 10-33cm(String Sized).Based on this the Grand Unification Energy GUT would explode at v=c while space-time exceeds it to accommodate the mass and energy. Many questions can be answered by the arc length equation

s=r

θ and this is a very simple relationship which can be utilized for the compactific

ation of type II a string theory to aid in the explanation of the BIGBANG.

If arc length or S and theta represent space-time and R or the radius can incorporate everything else it is possible to show space-time=space/massxc2 relates to s=r(theta).The N-branes where N approaches infinity=D-0-Branes+ or –Neven Schwarz 5-brane-1/2NS-NS 2 potentials associated with Fermionic Field intensity caused by the infinite momentum limit 1/R where the Fermionic Field is gravity or the curvature of space-time caused by R ab which is the Ricci Tensor associated with the metric g ab where the infinite momemtum limit is 1/R or infinity.As the D-0-branes=0 and the gravity space-time curvature metric is infinity when dealing with the infinite curvature of a point where the ferminonic field and bosonic field have 128 permutations each acting on 256 permutations of Riemannian or Mintkowski space-time you get 0/1/∞ at the infinite momentum limit where R approaches infinity you get 0/0 as 1Based on this is

ther1/∞ is zero therefore $\frac{0}{0}$ is mathematically everything including zero.

If the radius of s=rθ is $\frac{0}{0}$ or everything including zero and s and θ are space —
time there is an argument that this equation could be used   for the infinite dimension equatic

and to mathematically explain "The Big Bang" another way if the compactified form of IIa string theory is a circle.

CHAPTER NINE
HOW   WILL THIS UNIVERSE END?

There is a theory that this universe will end with "Heat Death" whereby the accelerated expansion with progressively decreasing rotation of space-time with the approximately 750 billion galaxies will slow down and eventually stop with the distances such that any gravitational effect caused by galaxies will become negligible and space-time will approach asymptotic flatness except for the effects of the Cosmologic Constant ^ suggestive of reciprocal curvature of space-time caused by Dark Energy almost cancels space-time curvature caused by the sum total of all the mass of each galaxy. When this happens eventually stars will cool down and collapse into black holes where the majority of gravity will be sequestered and the ambient temperature will approach 2.74 degrees kelvin resulting in widespread freezing.
 An alternative to this is the "popped balloon scenario"of space-time which may possibly trigger a massive implosion and a"Big Crunch". The speed 3x10^8 meters/sec or approximately light speed"c" is a boundary in which ordinary matter can't seem to breach. Tachyons as subatomic particles seem to be able to breach this boundary as well as going backwards in time(time's arrow reversed). This occurrence can occur if tachyons are bosons with a negative instead of positive mass which would make it impossible to tachyons to travel below the speed of light at which point they would be bosons traveling forward in time and with a positive mass. It has been postulated by this author and Steven Hawking that time's arrow can reverse in a massive implosion or "Big Crunch" or by this author that negative mass as may occur at the event horizon of black holes causing a superimposition of regions in Schwarzchild Space-time netting a solution to the Hawking Paradox as mentioned in a previous chapter. Normally in a Big Crunch which could have a slow phase followed by a pop inward of approximately Planck Time 10-43 there may just initially be a light Doppler shift toward the ultraviolet instead of the infrared showing the accelerated expansion of galaxies away from each other slowing .Note this can also happen in the early state of "Heat Death" therefore measurements would have to show a gravitational metric or curvature of space-time more than canceling the anti-gravitational effect of Dark Energy and the cosmologic constant to warrant consideration about a "Big Crunch" unless the "popped balloon scenario occurs in which case the implosion could be so fast that it would equal but be opposite the explosion of the "Big Bang" with gravity instead of anti-gravity predominating.
Space-time curves inwardly during the singularity of a "Big Crunch" as it would at the event horizon of a black hole .If a "Big Crunch" would occur the end of this space-time manifold could occur by a rip or tear in space-time as the expansion at a

progressively increasing rate and rotation at a progressively decreasing rate since "The Big Bang" .As previously mentioned the quantum bubble with infinite curvature space-time progressively uncoils towards asymptotic flatness .According to Kurt Godel's rotating universe which is shown by the clumpiness of the WOMP showing the BMR or baseline microwave radiation from "The Big Bang" the clumpiness shows increased uptake and diameter which indicates strong perturbations from the rotational vector of the expansion of the universe. Note that when you take an osculating plane that's rotating and expanding simultaneously and you take a cut of this plane the result is a spiral configuration which is what Albert Einstein postulated as the configuration of space-time in 1913.Note that in an increasing expansion with decreasing rotation space-time can achieve a rip or tear like an inflating pair of pants with a tear that becomes increasingly larger until it pops like a balloon which is a logical conclusion to Cosmic Inflation which is Dr .Alan Guth's 12 hypothesis for the expansion of space-time with conformal gravity. This tear occurs at "c" which is the boundary of the speed of light where dilated time acts as if it stops. The Lorenzian Transformation shows time dilated towards infinity or becomes infinitely long as "c" is approached closer and closer. This lack of space-time at the speed of light barrier reflects the small rip or tear in space-time with inertia being caused by the near infinite mass of the matter approaching the speed of light boundary and the resistance of space-time above light speed causes the rip which will in time increase in dimension until a "Big Crunch" s induced .During a "Big Crunch "slow phase time will slow down then stop and reverse. This only occurs if space-time travels at faster than light and created this boundary such that matter is pushing against a brick wall formed by space-time forming the near infinite inertia shown by the Lorenzian Transformation.It was also postulated that "the speed of gravity" is greater than "c" or the speed of light ,but as gravity is the curvature of space-mass caused by mass, gravity reflects the rate of curvature of space-time caused by the mass of the matter approaching the speed of light .As inertial mass for any matter approaching light speed approaches infinity the curvature of space-time at the speed of light boundary approaches infinite curvature or a point such as the tip of a cone at v=c where the cause is the pressure of space-time above v=c. If space-time exists at v is greater than c but shrinks to an infinite curvature point at v=c and then exists with the space-time curvature metric of Einstein at v <c; the speed of light barrier is a weakness of space-time that is similar to a small tear which can increase by the expansion of space-time.
It is much less likely that a "Big Crunch" will occur rather than "Heat Death" because the amount of mass versus space in this universe is very small except at the speed of light boundary or at the event horizons of black holes. If the number of black holes were to increase geometrically due to the implosion of many galaxies without those black holes "evaporating" the the amount of mass would increase compared to pinched off space-time at every event horizon. As it is there many be tears in the inflating balloon of space-time at each black hole event horizon. Despite this the space(3 space) vs mass(matter and antimatter) is so skewed toward empty 3 space (as compared with 4 space or space-time)that the curvature of space-time caused by the extant mass of all 750 billion galaxies +sum of all black holes wouldn't cause sufficient curvature of space-time to cause significant rips or tears ;and since the

ratio of 3 space to mass would increase as long as the acceleration of galaxies continue over time or in space-time the likelihood of a "Big Crunch" would decrease vs. Heat Death. If space were rife with black holes (active or dry)and if space-time was continuously collapsing in multiple spirals then the gravitational curvature metric would exceed that of Dark Energy and increase the chance of ripping of space-time unless all matter approaches 3x10^8meters/sec which is "c".

Of course with

8

$\frac{\pi G}{c4}$ *as the gravitational coupling constant approaching the cosmologic constant* ^

The chances of a Big Crunch would approach 50:50.That's 8(pi)G/c^4.It is highly unlikely that a singularity caused by non-natural causes will cause a "Big Crunch" with our level of knowledge and technology .As the Omega point(everything that is learnable had been learned)is reached ,the ability to either accidentally or deliberately cause a singularity would increase however with the Omega Point wisdom(knowledge plus experience)should prevent such an occurrence, although according to Quantum Mechanics there is a quantum state where a singularity will occur either accidentally or deliberately. This is delving on the realm of philosophy however.

As mentioned previously ,during "a Big Crunch" time will slow down ,stop or and reverse time's arrow as one goes from a greater to less entropic state reversing the Second Law of Thermodynamics during this singularity. Time's Arrow will reverse but it will still seems like it's going forward to everything and everybody within the system that the time is reversing in. It would be like saying that his body clock is synchronized for time's arrow to be pointing backwards not forward as Steven Hawking postulated happens in a Big Crunch and what this author agrees with .In this case time will point backwards until it reverts to time=0 at the point of the Big Crunch when space-time expresses infinite curvature as it does before The Big Bang and another Big Bang will follow with a heavily rotated component uncoiling in a swirl with a ballooning expansion as with cosmic inflation and the process will repeat. Again with "Heat Death"in the expansion also progressively decreases as does rotation until everything stops moving and everything freezes. This scenerio would be much more likely if there were no boundaries such as 0 degrees kelvin,"c"the speed of light boundary and spacelessness which can only be breached during singularieis. Sadly we will never live to know because the early stages of a "Big Crunch" may only be be shown by a Doppler Shift which is only slightly less to the infra-red and clocks even with gravity may not slow down a measurable degree because the measuring device is part of what's being measured and heavy masses will also dilate or slow time down which would skew the readings   .These phenomena follow the equation ℝa b c d=R a b c-1/2 R g ab/R ab (1/c^2)where space-time is curved inward. Space-time is pulled outward by inflation or with H(the Hubble expansion coefficient)with slight rotation leading to a possible swiss cheese effect in our space-time fabric in which according to Guth and others 14 other universes can form with the same or different physical laws ;however the WOMP doesn't clearly show a swiss cheese effect in space's BMR. In this cast the equation of everything becomes $\mathbb{R}$ a b c d=R a b c+1/2R g ab/R a b times $1/c^2$ instead of -1/2 R g a b because space-time is curved outward rather than inward. Also it is an axiom that each and every intelligent observer must rust his or her perception of the information gathered by the measuring device and that each measurement is based on the Heisenberg Uncetainty Principle which state measurement ranges for everything measured as the measuring device affects what is measured changing it. The is also the information exchanges of relating to "Spooky Action at a Distance"and quantum states or levels changing polarity or information at vast distances. It is also an assumption that the observer exists.

CHAPTER TEN

WHAT OCCURRED BEFORE THE BIG BANG AND WHAT CAUSED IT?

The accepted theory was that time and space began at 10-43 seconds before "The Big Bang" also known as the "Big Bang ex nihilo". This theory   opposes Newton's Law "Every action must have an equal but opposite reaction as well as "The Law of Conservation of Energy" which opposes the spontaneous appearance of a quantum bubble out of nothing (including space).

It seems logical that an implosion of another universe could have occurred prior to the explosion of the "Big Bang" in which times arrow was reversed during the implosion of the previous "Big Crunch "at which time instant time almost stopped then reversed time's arrow from backwards to forwards.
There may have been a massive Higgs Energy field which converted a quantum bubble at 10-^33com to matter according to be variation of energy=mass c^2 and D-0-Branes may have also existed where the zero dimensional state approximated the infinite dimensional state where the infinite dimensions were compactified or curled up with a quantum foam or something else. The multi-verse idea 15 with a n implosion preceding the explosion of "The Big Bang" seems possible.

 It is also possible that three universes collided where one was predominantly matter ,one predominantly anti-matter and and one a vacuum universe containing energy only. This may be an unlikely event however quantum mechanics states that all permutations will occur including this one. In the anti-matter universe time moves forward and with matter time moves backwards .In the matter universe time moves forward   due to positive mass and antimatter time moves backwards due to negative mass. In the "vacuum universe "there is space due to the energy of possibly the Higgs Field and has mass due to the Higgs Bosons which are massive and almost equal matter and anti-matter   ,however there is slightly more matter than antimatter. As a result of the three way collision   ,the energy of the Higgs Field of the "Vacuum universe "triggers the implosion of the matter and anti-matter universes triggering a reversal of times arrow and imploding the mass of the antimatter and matter down to a quantum bubble which is approximately string sized which will subsequently explode 10^-43 seconds later into "The Big Bang" with initially high rotational vector at maximum magnitude just at Planck's Time then slowing at a geometric progression while the expansion or inflation occur with "The Big Bang" moving the rotating 750 billion strings into approximately 6.75x10^34 erg of energy in a 360 degree of

2

πRadian orb blast with a rotational vector ω and the Hubble Expansion CoeffiicentH^.

Collisions of galaxies occur infrequently but they do occur ,as there is a theory that Andromeda is approaching The Milky Way for a collision in several billion or perhaps many many million years ,so the collision of universes can and must occur also therefore the term "Universe" is a misnomer but is rather a "Multiverse". Matter universes exist ,anti-matter universes must exist because antimatter exists. Pure energy universes must exist with the Higgs Field because the Higgs Field exists. As all of these axioms are true the collision can occur and must occur at time "t" though highly infrequent. An since this scenario will cause the quantum bubble to occur and since a vacuum universe will force an implosion rather an explosion, this hypothesis seems very likely. Also the area   between membranes of universes has to exist it can be called or referred to as "ether" ,which would include D-0-branes which merge with n-branes in a matrix of Quantum Dots. This explains "The Big Bang" or swirl and what precedes it where the Big Bang isn't out of nothing.

From this author's first book "Megaphysics,A New Look at the Universe"it was postulated that two two anisotropic manifolds of ten dimensions each had a singularity with a dimensional reconfiguration of two six-dimensional manifolds and two four dimensional manifolds which resulted in a twetnty dimensional manifold of different configuration than the two initial anisotropic manifolds or surfaces which were unstable. With regard to theories regarding the origin of this universe Michio Kaku's idea that a 10 dimensional syupersymmetric anisotropic universe had a "popped bed sheet" effect with the popped bedsheet being the six dimensional Calabi Yau Manifold which was string sized and the four dimensional superstring cosmic universe made more sense than the "Big Bang ex nihilo".15In my theory that alternates from the three way collision theory there were two ten dimensional manifolds totaling twenty dimensions and a six dimensional Calabi Yau manifoldor something else possibly related to the Higgs Boson with the Higgs Field acting as "ether "cohabitating a "false vacuum" containing infinite space and energy from the Higgs Field. In this scenario a Higgs' Field Universe ,matter universe and antimatter universe didn't have a three way collision nor were there an almost infinite number of multiverses ,although there still were an infinite number of dimensions primarily forming D-0-branes where the dimensions were compactified from infinity to zero in a quantum foam or quantum dots in the "ether" with the two Calabi Yau manifolds and two 4 space superstring manifolds forming twenty dimensions with a huge six dimensional Calabi Yau manifold forming the other six dimensions of the non-compactified 26 dimensions of string theory cohabitating the quantum dots of D-0-branes.

The twenty six dimensions indicated by string theory included Ramajian's magic number of nature of 24 as the most stable state of four Calabi Yau Manifolds on the dimension of time and another dimension involving the Higgs Field which incorporates into the other set of four six dimensional Calabi Yau manifolds with time moving backwards or forward depending on the properties of non-compactified space and their near infinite dimensional non-compactified quantum dots. An absolute vacuum developed in an infinitely short period of time(time approaches zero but doesn't reach it)and a tidal wave of near infinite space region engulfed space-time and its contents creating a huge amount of vacuum energy from a singularity according to the equation E=mc^2 which acted as a PRIMER or catalyst either for the Higgs Field to act on the three way collision of the matter, antimatter and Higgs Field Universe to form the quantum bubble or the mass formed torn or ripped the symmetry of the recombined two ten dimensional super symmteric anisotropic universes to form an antimatter quantum bubble a matter quantum bubble both 10-33 cm with one 12 dimensional forming two Calabi Yau manifolds and two 8 dimensional forming two quantum bubbles which each contain 4 space with opposite spin2 vectors ;one clockwise one counterclockwise forming the rotational vectors of the initial 10-43 seconds before the "Big Bang" or swirl. The eight dimensional string bubbles had a crushing Ricci Tensor which was opposed by the anti-particle anti-particle repulsion twisting it into a double torus configuration which is more stable than the 8 dimensional complex causing rotation to occur greater and greater at the ends of the bubble(s)where the rotations were in opposite directions(differential spin)until the double torus acted like a pretzel and finally divided in in the middle of the two four dimensional quantum bubbles rotating in opposite directions with most of the mass occurring at the ends or outer edges and finally reaching a velocity to exceed quantum gravity of the two more stable isotropic four dimensional universes leaving two 6 dimensional string universes .The bulk of the mass spun out like a potter's wheel toward infinite length but remained string sized with Calabi Yau Manifolds at 10-33 cm. Whichever of these ideas is the simplest according to Ockham's Razor (Occum's Razor) is the most likely but again with quantum mechanics there is a probability that both occurred .In any case,"The Big Bang ex nihilio "doesn't seem likely in terms of Newton's Third Law and The Law of Conservation of Energy and Momentum.

CHAPTER ELEVEN

DOES NOTHING EXIST?

By definition nothing means non-existance.Mathematically space-time/space-time is described by the equation -1/2e^in

$cot\theta\, divided\; by\; -$

$\frac{1}{2e^{i}n}\cot\theta.\; In\; the\; expression\; i\, n\, \cot theta\; where\; theta\; is\; \pi\; radians\; you\; get\; i\infty cot\theta\; or\; -$

$1(i\infty)if\; \theta = \pi\, radians\; -\frac{1}{2e^{\infty}for}n\; dimensions\; where\; n \rightarrow$

$\infty - \frac{1}{2}e\; to\; the\; infinity\; power\; is\; =$

∞. *So as space* $-$ *time is* $-\frac{1}{2e} -$
i n cotπ is ∞. *The expression* $\frac{\infty}{\infty}$ *is everything except* 0(*zero*). *So as the* $-$
$\frac{1}{2's}$ *cancel out e* $-$
i n cot θ where n $\rightarrow$
∞ *gives* ∞ *in the numerator and denominator plus THE IDENTITIY POSTULATE IS MET SO WI*

INFINITE SPACE-TIME IF THERE ARE INFINITE DIMENSIONS APPROACHED IN THE MULTIVERSE.As space-time is never stationary except in a spaceless absolute vacuum due to space-time's motion(eventhough it goes below Planck Length at the Event Horizon of a black hole or pre-Big Bang or post Big Crunch)the infinite sum of space-time ∞ *and must be* ∞. Π(*eigenstates* 0 *to* ∞ *of a1* $\longrightarrow$ *an where a =*
space $-$
time diverges to ∞. $\frac{Again\infty}{\infty}$ *is everything except* 0. *If the expression was* $\frac{0}{0}$ *then* 0 *would have bee*

Included. But as zero spacetime is not the infinite sum of spacetime the former MUST BE THE CASE.The exception would be if all space-time moves in equal and opposite direction with the same magnitude always..As the Riemann Metric for space-time with its 256 permutations equal zero net and the Bianchi Identity cancels out space-time in equal and opposite directions in many covariant and contravariant tensors there is an argument that spacetime(net infinite sum)/spacetime(net infinite sum) is 0/0 which includes the null set or spacelessness.This is true as 0/0 includes 0 as a solution.As the meat of the question is does all space-time as an infinite sum of the metric of R a b c d equal zero.According to Riemann it does. This indicate the mirror space-time manifolds of -1/2 e ^I n cot theta and -1/2 e –in cot theta as the left and right handed components of space-time in the Big Swirl and Anti-Swirl mentioned in this author's first book "Megaphysics ,A New Look at the Universe".Ths is true in the n-dimensional state where n approaches infinity with regard to dimensions. Although space is a container by definition and a container can exist without contents this seems to go contrary to logic although Einstein apparently stated that mass creates space but this precluded "massless "energy fields with photons which while they have a positive moving mass have a zero resting mass according to sources. Spacelessness is a boundary which if breached can implode a universe and possibly the entire space-time continuum reverted to a Higgs Field of background energy and quantum dots which are infinite non-compactified dimensions in a strict lattice formation as described by the 2 dimensional lattice equation based on the critical exponent and the act of bosons on a circle which is a compactified form of type IIa string theory or M Theory.The spherical model 17 if solvable in the presence of a field such as the Higgs Field .The spin takes on real and imaginary values and interacts with all the spins of the quantum dot lattice .It is subject to the constraint $\Sigma a = 1$ *to N* σ^2*where* σ *ais the first eigenstate of the sum of sigma squared =*
N. Couples that with the thermodynamic limitof G(Gibbs Free Energy) =

$$-kT/$$
$$2\pi \int_0^\infty F(\theta)d\theta \text{ based on the Boltzman Equation of States of matter interacting with energy}$$

and theta becomes infinitely small we develop from the two dimensional Ising
Model a two dimensional statistical model where a paritition function
Z=

$$\Sigma n \, e\left[-\frac{E(n)}{kT}\right\} \text{ where } E(n) \text{ is the energy of the nth state k is the Boltzman constant and T represe}$$

the temperature down to zero degrees kelvin. The free energy F=-kT ln Z at
criticality correlation functions between spins

$$\sigma i \text{ and} \sigma j \text{ develop a metric } g\,ij. \text{ The expression } g\,ij =< \sigma i\sigma j > -< \sigma i >< \sigma j >$$
$$\text{depending on the distance } x \text{ or } r \text{ spearating the states. The correlation length } \mathcal{E} \text{ becomes } \infty$$

At a phase transition from one state to another and at large distances where x or r
approaches infinity g i j approaches x-te-x/ε $(.\Delta = conformal\ weight) Thus\ g\ ij =$
$$x^d + 2 - \eta =$$
$x^{-2\Delta} where\ these\ are\ approximations. While\ \eta is\ the\ critical\ exponent of\ the\ field\ the\ energy\ op$
16.Phase transitions on a quantum dot level approach the Ising Spherical Model 18
energy
operator ϵ as a product ogf two fields whereby $\epsilon\,n =$
$\sigma n\sigma n + 1$ then it follows that $< \epsilon n\epsilon 0 >$

equals $x(any\ observable\ or\ expectation\ value$ $^{-2\left(d-\frac{1}{v}\right)}$ has an infinite correlation length at a
at a critical temperature which was already mentioned. The interaction of a free
boson or Higgs Field on a quantum dot matrix may follow the Ising Spherical Model
as the circle is compactified type II a sting theory. If spaceless ness occurs a fracture
will occur in the quantum dot lattice upsetting the
spins
$\sigma 2^2 breaking\ the\ continuum\ N\ relating\ to\ the\ sum\ of\ spins. This\ state\ is\ in\ the\ zero\ external$

or fermionic (vacuum state).This is shown by the Grassmann Oddvariable
$\psi - n^2|0> with\ the\ nth\ Fermionic\ Oscillator\ trace\ being\ \psi - n|0 > and\ vacuum\ state$.

This describes the quantum dot lattice in terms of two dimensional open or closed
strings using the Ising Model 19. As the Fermionic vacuum state and quantum dots
are not nothing and the spacelessness state actually can be subdivided down toward
infinity where an infinite number of dimensions equals zero dimensions as in the D-
0-Brane spacelessness doesn't exist.

A NEW DERIVATION OF SPACETIME=SPACE/MASS

In terms of metric tensors space-time is curved Lorenzian Space-time or R(region

of topological space acting upon or being acted upon by the tensors of the 256

permutations of space-time in tensor of the fourth degree represented as R abcd

with the different permutations of the different eigen-state of energy from the

ground state or the cosmologic constant to 10^77 joules tying in with the energy

associated with the Higgs Field and the Grand Unification Energy of 10^19 giga

electron volts with recombination from each energy level commixing with each

other energy level.this is entanglement.

$$\Gamma R\,abcd(1/\sqrt{n} \quad 10\text{\textasciicircum}77 \text{ joules}|\text{Cosmologic Constant}||1 \ 0 >$$
$$\mu \; = \; \Lambda + or - \; R\,abc + or - \tfrac{1}{2} R\,g^{\underline{ab}} \div \; i\,\hbar\,\rho\,abc$$

1. Here the derivative of the tensor of the fourth degree with entanglement

    over n eigenstates of energy grom the cosmologic constant to the Grand

    Unification Energy or 10^77 joules or both is spacetime and this equals the

    cosmologic constant plus Euclidian flat space + antigravity(1/2) –

gravity(1.2) divided by the square root of -1 times Planck's Constant/2pi

times the energy density of matter rho over abc the tensors of Euclidian flat

space where the denominator equals R ab(Ricci Tensor reflcting inertial

mass) or R ab=i$\hbar\rho$ abc where h (bar) = $h/2\pi$

DESCRIPTION OF THE HIGGS FIELD WITH INTERACTIONS

 Describing "The God Field or Particle "has been difficult in attempting to fuse, science, philosophy and religion. The definition of "catholic"is universal;therefore cathalocism is the religion of the universe. Based oon this there must be new definitions.

It is indeed difficult to describe the Higgs Field in totality as some assumptions will be difficult for most of the scientific community to "swallow" ,however as the Higgs 'Boson or Higgs' Field describes the "body" of "God" which can convert energy into matter and back into energy. The semi-permanent hemi-radius tachyon has always existed and will always exist as we perceive time in terms of relatively constricted time and dilated space(a relative vacuum space) The point of creation which was the Time Oscillation Paradox had a rogue tachyon or tachyons drop to the speed of light which is a relative boundary(soft vs.hard boundary).Note:The only hard boundary is that of spacelessness or nothing;which doesn't exist).Was this drop of the rogue tachyon deterministic or random? If it was a group or grouping of tachyons going backwards in time to encounter leptons and gluons turning into Bosons(which go forward in time);then the event would be deterministic and there would be an argument that tachyons are the brain and nervous system of the Higgs'Field. Incorporating an idea first broached by Veneziano in 1993 that this universe is "alive"(at first though of as a preposterous idea but with Quantum Mechanics not so preposterous)one can describe the Higgs'Field(God?)more accurately. When time dilates it constricts space;when time constricts it dilates space,which is why traveling into an infinite space vacuum propels one into the future. In constricted time,time moves almost infinitely fast into the future(times' arrow forward .In dilated time ,time moves almost infinitely slow relative to the observer giving flat curvature to no curvature in infinite flat space and space is constricted by the dilated time into an infinite curvature point or dimple in perfect fluid space-time.
Galaxies rotate around a central black hole and rotate very slowly relative to the observer.Therefore they rotate in dilated time relative to us much as the event horizon of any active black hole exists in dilated time and constricted space(Schwarzchild Space-time)relative to the observer..Think of time as wrapping around space and constricting it much as a rope strangles a ball until the ball is constricted to a point if the rope is infinitely long.If the rate of rotation of each and every galaxy is perceived as faster it can be considered a dynamo producing energy much as mitochondria does in a unicellular organism. Black holes can be like pores in skin or an excretory function eliminating "waste material "and "The Big Bang" can be considered like fertilization. If a unicellular organism or a lymphocyte is able to conceive an entire body it would be analogous to the observer conceiving the Higgs'Field.
Coupling these lofty ideas with tachyons acting as a brain and spinal cord to the body of the Higgs'Field acquiring energy from rotating galaxies and excreting from black holes one can determine THE MEANING OF LIFE. As a white blood cell in our

bodies contributes to the well being of our bodies we would contribute to the well being of the massive Higgs' Field and Higgs'Boson with tachyons acting as the brain and spinal cord and with "birth "being the" Big Bang" which was not the first event as described earilier.As fertilization would not be the First Event but a birth in a continuum of universes;the multi-verse or a collection of Higgs'Fields with tachyons acting as the brain and spinal cord. This concept clearly fuses religion,philosophy and science.

ENTANGLEMENT;HOW FAR DOES IT GO?

Lately ,new scientific evidence(Science News;Emily Conover vol.191 #8

4/29/2017) indicates that at least millions of bits of information on a sub-molecular

level become entangled with each other. This has far reaching implementations as

entangled particles with data can mix or exchange information of any type

anywhere.

Can this be associated with paranormal phenomena?" Mystics" who via séances can

purportedly contact the dead may in a few select cases have their frontal and

temporal lobes of their brain entangled with data from deceased relatives or other

individuals. In my previous book "Mega-physics III; Nothing Doesn't Exist "it was

explained by this author that "Spooky Action at a Distance" can show information

exchange with quantum entanglement between an evaporating black hole and

adjacent black holes or even black holes in other galaxies in milliseconds thus

solving "The Hawking Paradox". Can ghost sitiings be related to entanglement? If

memories or images from the past are stored in the hippocampus of the brain of an

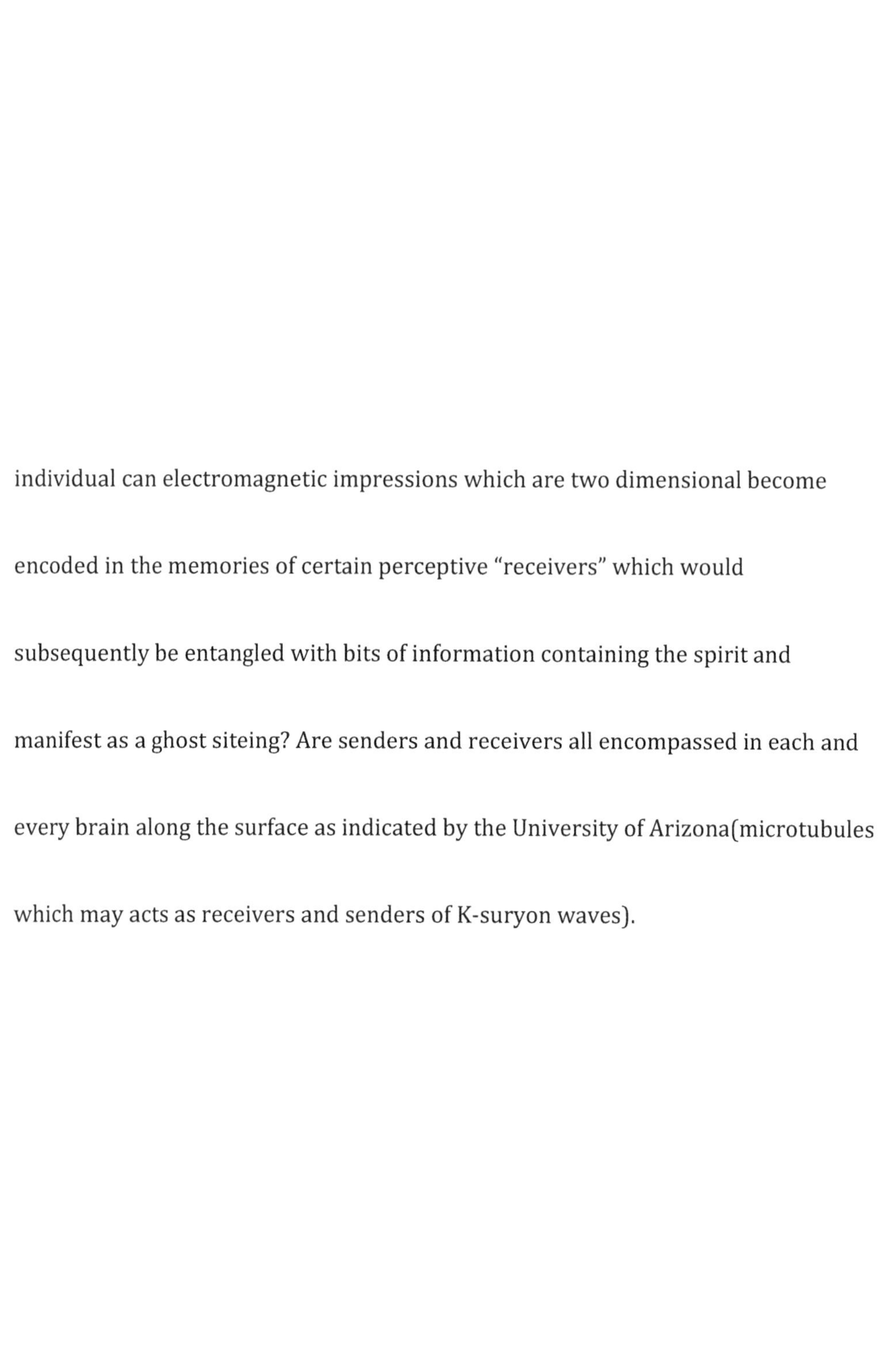

individual can electromagnetic impressions which are two dimensional become

encoded in the memories of certain perceptive "receivers" which would

subsequently be entangled with bits of information containing the spirit and

manifest as a ghost siteing? Are senders and receivers all encompassed in each and

every brain along the surface as indicated by the University of Arizona(microtubules

which may acts as receivers and senders of K-suryon waves).

Entanglement of time or the sequencing of events can cause the unexplained

phenomena of "bi-location" whereby a person or animal can be in two places at once

at great spatial distances. If a person or any animal ,plant ,fungus,bacterium or virus

can exist in two or more separate times in the same space(rather than the usual

different spaces)it could be as in "Spooky Action at a Distance "be displaced in two

different time elements due to a Time Oscillation Paradox with locality being shifted

to a non-local event. The aetheric double as it's called could be the same individual

shifted in time either in the past or future with entanglement being the time

oscillation which must occur in time travel backwards however a complement or

recpiprocal of time travel forwards. To appear in two separate times but within the

same space indicates that space-time coils and uncoils in predicted pattern

interacting with other space-time in a non-predictable or random manner making

the aetheric double appear randomly elsewhere.To make it deterministic would have K-suryon waves interact with coiling or uncoiling space-time to cause a

deterministic time oscillation and forced entanglement to another time and place

explaining bi-location.

HOW DOES STRING THEORY AND M THEORY PLAY INTO THE EQUATION OF EVERYTHING?

String theory is based on the mathematical concept of how notes and octaves on

musical instruments follow certain patterns that are not only present in our

universe and other universes but actually dictate actions of metrics(any measurable

entity).   There are five different string theories which have duality to each other.

Type I; which describes the behavior of strings which are open and connect or

disconnect from each other twist and turn in multiple dimensions and dictate

nature, Type II(which are closed strings that twist and turn andi interact with each

other in multiple dimensions and break with reformation .Type IIA which is similar

to type one but adds modular theory to make the strings C.P.T. variant by having

closed and open strings interacting with each other in multiple dimensions SO32

which in Yang Mills Theory is an energy level tied in with open and closed strings

with string open in 32 manifolds dictating 26 dimensions compactifiying to 10. The

fifth string theory is the Heterotic **8x8 which involves the flipping between the**

different dimensions of open and closed strings as grouping of an 8x8 matrix

or determinant as described in Tensor Calculus. These five string theories are

dual to each other as they observe the same phenomena from different

vantage points such as observing a car from the front rear side and above.

Utilizing string theory with the Tension of each string as being massive 10^32

tons/string varying down to 10^39 tons depending on it's interactive

relationship with both space-time(the orbifold in string theory) and other

strings at different eigen-states of energy. The formula mass of a string

squared=2pi Tension on a string(number of eigen-states of energy)describes

the Tension of each string which would increase geometrically with an

increase in the energy state as would the motion of each string in each of the

26 dimensions to interact with other strings in the Plasma State of matter

comingled with kinetic energy reflecting as heat and high energy radiation

such as gamma rays and x rays. Inside a black hole past the event horizon the

pressure of the strings interacting with each other would exceed 10^39tons

and may be even greater . The temperature can be determined by hawking's

formula for the temperature of any active black hole(which now is believed to increase in size and mass as the accelerated expansion of space-time in our

universe increases with infinite Dark Energy.

S=rθ the equation of arc length when applied to the osculating plane will pptro an infinite number of dimensions. The unit tangent vector of a sphere of

2

$\pi radians\ in\ motion\ is\ 2\ \pi\ R\ cos\theta\ where\ \theta\ \ is\ \ \pi\ radians. The\ sphere\ travels\ at\ HOR\ THE\ HUBBLE$ EXPANSION FACTOR.$As\ \theta\ approaches\ zero\ \cos 0\ appraches\ 1\ causing\ space-$

$time\ to\ approach-\dfrac{1}{2e}-i\ times\ the\ number\ of\ dimensions\ \ where\ e=$

$2.71828\ then\ spacetime=$

$-\dfrac{1}{2}e-$

$i(to\ the\ n\ power)x1\ where\ n\ approach\ \dfrac{es\infty\frac{or1}{-1}}{2ei\infty}\ power\ or\ 0\ which\ indicates\ that\ n\ infinite\ number$

Down to theta approaches zero causes r or the arc length at the circumference to approach zero

CONTINUATION OF MATHEMATICAL EXPOSITION

R ijkl -   R ji=-8πG e (ij, ji) =g
(kl above; ji below)

-8πGe(ij, ji)=g ji
  vector product of Rij 'Rji=e(ij, ji) cos   π =-e(ij, ji)
   8πG = Λ   -1 is from cos θ   θ = π   klR ij.R ji ji for space-time
curvature from antiparticle of metric g ji on antiparticles   Rji kl=contra-variant
tensor kl on covariant tensor ji ji is from antiparticle g ji kl is from gravity of
contra-variant tensor for metric g ij for matter
   Ijkl=antigravity effect on particles from antiparticles

R ij Rji (squared)
--. e ji where Rij Rji/$\left\|R^{ij}\right\|^2\left\|R^{ji}\right\|$ =cos θ

$\left\|Rij\right\|^2$ $\left\|Rji\right\|$

add e kl to e ij e kl= R kl /$\left\|R kl\quad R kl\right\|^2$ 'g kl   cosθ 'g kl=-1 . g kl whereθ =
π radians   as covariant tensor For contravariant tensor e kl = R kl squared/
$\left\|R kl\quad R kl\ .g-g kl = g lk\right\|$

R ijkl = g kl/g ij'gkl=-
(kl above R)
$8\frac{\pi G}{-16\pi G} = \frac{1}{8\pi G}$ whioch is the reciprocal of the cosmologic constant Λ 8πG =
$6.67x10\frac{11}{8\pi} =$
7.5x10 10 joules −
seconds which is a huge antigravity effect of Dark Energy from the antimatter antimatter e⟩

The gravitational coupling constant k=-8πG/c4   -8πG/c 4 is also synonymous
with antigravity between 2 particles of matter or anti-matter .As matter- matter
interactions have a positive gravity   anti-matter anti-matter interactions are
antigravity Q.E,D.

Using E=mc2 with the number of strings in the quantum bubble for antimatter and
matter being the mass the calculation emerges as E=7.5x 10 10   joules(3x10 10
cm/sec)2=6.75x10 31 joule-sec or on one second 6.75 x10 31 joules as the blast
force from "The Big Bang" ;as antigravity is the predominant blast force Dark Energy
blasted out in the 360 degree Orb Blast producing 750 billion galaxies from the Mas
string calculation. Note 10 31 is 10 to the thirty-first power

                                                                                    4
NOTE: Although some sources   list the cosmologic constant (∧) as 8πG/c   this
has no impact on the mathematical calculations only labeling. Other sources list
this as the gravitational coupling constant.
G is in kg-meters/sec 2

6.67x10 11/8π =7.5 x 10 10 for   number of strings. Calculation or at 10 7 erg per
joule for 'Big Bang' blast force is e=mc 2 or e(joules)=7.5x10 10(3 x 10 10)2=75 x
10 30 joules or at 10 7 erg per joule 75 x 10 37 erg or 7.5 x 10 38 erg

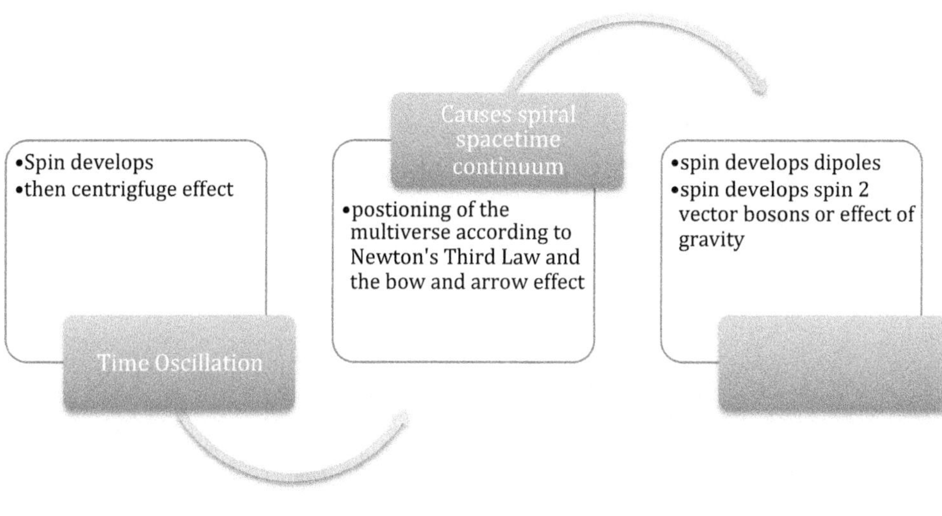

•Spin develops
•then centrigfuge effect
Time Oscillation
Causes spiral spacetime continuum
•postioning of the multiverse according to Newton's Third Law and the bow and arrow effect
•spin develops dipoles
•spin develops spin 2 vector bosons or effect of gravity

tachyons above the
speed of light
moving backwards in
time
time oscillation with
fermions traveling
forward in time below
the speed of light

APPENDIX

BRANES;A STRING IS A one-BRANE WHICH COUPLES TO A BACKGROUND SECOND DEGREE TENSOR.Zero-branes are ten dimensional building blocks for space in the pre-Big Bang epoch.The second degree tensor is purported of negligible mass as indicated by the ZERO-BRANE.THE SOURCE OF THE BACKGROUND SECOND DEGREE TENSOR IS R uv where the integral of D to the d power of x where x is the string or one-brane in D dimensions applies to R u v g u v where g u v is the metric acting on R u v for the zero-brane with respect to x which is the one-brane. R u v is the second degree tensor upon which the metric g u v acts. In four dimensions a monopole is dual to two electrons acting on a zero-brane. In 10 dimensions a string is analogous to a five-brane based on p-brane potentials. This involves dual fields such as a tensor R=R* from Ra1...a n=R8b1...b n.p-branes are encircled by a hypersphere which relates to M theory being compactified (curled up)by a circle for typeIIa strings.The charge of a p-brane is based on Q=∫ $\quad *$
$R \; from \; limit \; S \; d - p -$

$2 \; for \; electric \; charges \; and \; Q \int_{Sp+2} R \; for \; electromagnetism.. P -$
$branes \; tie \; in \; with \; the \; potential \; involved \; with \; permutations \; of \; a \; field \; tensor. p \; brane \; tensors \; a$

Are associated with a tensor of the pth rank R a1...a p and electric and magnetic charges can be associated with p-branes with superalgebra.

Dzero-branes represent the vacuum state.ALTHOUGH INDICATED AS ten dimensional building blocks of space they actually are zero-dimensional. ONE-BRANES REPRESENT STRINGS WHICH

ARE TWO DIMENSIONAL OR POSSIBLY ONE DIMENSIONAL.If all the dimensions in a system or universe are conserved such that the total number of dimensions are constant;then zero branes would have to be ten dimensional in the vacuum state.Six dimensions for CalabiYau Manifolds and four dimensions of space-time. As the "c" boundary is approached infinite mass with reducing length and width occur when length becoming infinite.In this case width and height approach zero but do not reach it and become infinitely small curled up and compactiifed.In general although showing duality between different systems which are abelian membranes are described by the forces involved with mass or energy associated with the membrane with reference of n-dimensional space where n dimensions would have n-1membranes or n-1 brane.

FLAT OR MINTKOWSKI SPACE IS DESCRIBED MATHEMATICALLY AS THE LINE ELEMENT OR ds2=dx2+dy2+dz2-c2dt2.FLAT SPACE-TIME IS SPACE-TIME WITHOUT ANY CURVATURE IN OTHER WORDS A VACUUM STATE HERE R g a b=0 which indicates that the space-time curvature metric=0 and therefore gravity =0 in the vacuum state.
CURVED SPACE-TIME IS GENERALLY DESCRIBED BY ds2=e-k|r|(dx2+dy2+dz2-c2dt2)+dr2 where r=space-time curvature metric described by tensor as R g a b.R g a b or r is determined by the inertial mass of the object doing the curving and the curving is performed by bosons and possibly gravitons or fermions.Spiral space-time has k=-i(the square root of-1)to the n cotangent theta power as suggested by Dr.Roger Penrose and proposed by this author.

MANIFOLDS
THE SIMPLEST MANIFOLDS ARE CARTESIAN SPACES WHERE A MANIFOLD STRUCTURE OR SURFACE IN TERMS OF TOPOLOGIES IS R to the d power with what's called an identity map Rd implies R d .The coordinate functions of this map are cartesian coordinates.If coordinates are a I ;R d is the manifold of the standard Cartesian coordinates.a i=ax+ay+az and R to the d power is the tolological expression of the standard manifold or

the Cartesian Coordinate system. If a manifold is imbedded in another manifold it is a submanifold.On a string basis submanifolds can be orbifolds or Calabi Yau manifolds which are submanifolds for spiral manifolds for asymptotically flat but curved space-time on a macrostatic surface which is expanding and simultaneous rotating as at black hole event horizon. RIEMANNIAN CURVATURE A RIEMANNIAN SPACE IS THE SPACE COORDINIZED BY xi(power)with a fundamental form of the Riemannian Metric g I jdx I dx j where g=(g ij) obeys the metric tensor.g is of differentiability class C2(all second order partial derivatives of g I j exist and are continuous.g is symmetric g I j=g ji;g is nonsingular |g I j| doesn't equal 0.The differential form and distance from g isn't variant with regard to changes in coordinates.

R I j k l=g I ir(Rr superscript with jkl as a subscriptwhere R jkl with ias a superscript is the Riemann tensor of the second kind.The Riemann Tensor of the first kind is R I j k l=$\frac{\Gamma jki}{xk} - \frac{\partial \Gamma jki}{xi} +$ $\Gamma ilr \Gamma jk$ with r as a superscript $+ \Gamma ilr \Gamma jk$ with r as a superscript $- \Gamma ikr \Gamma ji$ with r as a superscript.**Here** Γ is a Christoffel symbol or the derivatire of a tensor. Above is $Rijkl - \frac{\partial \Gamma jli}{\partial xk} - \frac{\partial \Gamma jkl}{dxl} + \Gamma ilr \Gamma jk$ with r as superscript $- \Gamma ikr \Gamma jl$ withr as superscript. Skew Symmetrys involve Bianchi'sIdentity $R ijkl + Riklj + Riljk = 0$ skew symmetry is $R i j k l = -Ri j kl$ an d second skew symmtery is $R ijkl = -R i j l k$ with $R j k l$ with i as superscript $= -R j l k$ withi as a superscript. Block symmetry is $R i j k l = R k l i j$. These symmetry properties must fir with the $n2(n2 - \frac{10}{12} COMPONENTS OF THE RIEMANN TENSOR (R i j k l)$ where the diagnal tensor without s. PTR

.

Rijkl=g I iRjkl is subscript and I as superscript in the diagonal metric te tensor calculations for the Riemann Metricgives six cases R one R 212 and 1 R 313 and 1 R 323 and 1 R 213 and 1 R 232 and 1 R 123 and 1 which proves with the partial derivatives of the Christoffel symbols of tensors according to the previous formulas give R I j k l=0 for all I j k and I indicatin the summation of all Riemann forces and space is zero.The math of all these combinations is very difficult to reproduce by typing.

GLOSSARY

Abelian :equations having a coefficient or variety in a specific group,g,,algebraic number fields,tensors of the same degree or cohominy group

Anisotropic :not isotropic,lacking observational symmetyry

Anti-symmetric:tensors or vectors that are equal but opposite and can therefore partially cancel or cancel

Aymptotic:that which approaches a level or degree but never reaches it;asymptotic flatness appears without curvature but doesn't reach it

Bianchi's Identity: The identity of groups of Riemannian 4 space that is anti-symmetric and Abelian and cancels each other out of being equal but opposite
"The Big Bang"A theory proposed describing a Friendman type I open expanding f;at universe with is homogeneous and isotropic

"The Big Swirl "A Big Bang with a progressively decreasing rotational vector from an infinite curvature point of space-time to asymptotic flattness

Black Hole :collapsed matter from a neutron star or galaxy with extreme curvature of space-time at the central nexus due to extreme gravity of of the spiral space-time

Calabi Yau Manifold:a surface which represents a relative isotropic portion of space-time with a puckering to accommodate multiple dimensions considered a twisted variant of the orbifold
Choas:absolute disorder

Chiral:a mirror image or absolute symmetry

Closed string:a two or one dimensional building block of matter from energywith movements in 10 or 26 dimensions without breaking the string

Compactified:when every point of the dimensions are curled up mathematically making the size approach zero.First determined by Kaluza and Klein

Conformal Space:when every point in space relative to every other point maintains its relative position regardless of what the space is doing

Dark Matter;an indirectly measured mass causing perturbations in gravity(the curvature of space-time)caused by mass.Acts as cosmic glue containing possibly baryonic particles and neutrinos

Event Horizon:area where a black hole is perceived by measurementsEntropy:degree of disorder

Entropy:degree of disorder
Ex nihilo:out of nothing

M(Membrane)theory:the 5 dual string theories into one massive theory of everything which incorporates membranes which vibrate and incorporate all energy and matter

Isoropic:observational symmetry

Geodesic:a unit of space-time
Gravity:the curvature of space-time caused by mass;actually an effect not a force

Membranes:a description of matter in terms of energy states with stress energy densities described in the number of states with regard to dimensions

N;number of dimensions in N dimensional space

Open string:a two or one dimensional bulding block of matter with movements in a multidimensional plane

Orbifold:space-time manifold in an open twisted cone configuration utilized in string theory

Relativity:the behavior of matter and energy with regard to other matter and energy;energy and space have a different vantage point from other matter and energy including stress energy,time and mass with changes regarding relative velocity

RicciTensor:that tensor which represents inertial mass or resistance against pull or push
Riemann Forces:all strong and weak forces in nature

Riemannian Space:Mintowski space with Riemann curvature of space-time caused by mass.Flat space if no mass is present

Scalar:the magnitude compone t of a vector or tensor with regard to direction

Six dimensional string manifold:curled up closed strings in configuration according to Kaluza and Klein which is 10-33 cm and may be Calabi Yau Manifolds

BIBLIOGRAPHY

1:Peat,F.David. Superstrings and the Search for the Theory of Everything.Yang Mills Forces p.114

2.Kaku,Michio. Strings,Conformal Fields and M theory.Ising Model p.176-78
3 : Wald ,Robert .General Relativity.Chicago,Ill. University of Chicago Press.1984 4.CPT THEOREM; Quantum Field Theory Kaku ,Michio
4:Metric tensor(General Relativity)Wikipediaa and Spacetime.en.m.wikipedia.org.spiral Space-time Einstein 1912 Fractal Time.p.108-109 Braden ,Gregg 2009 Library of Congress HAWKING RADIATION. Wikipedia
5.Peat,F.David .Superstirngs and the Search for the Theory of Everything.p.106-107.Calabi Yau Manifolds

6.Kaku,Michio .Quantum Field Theory. Renormalization Actions in Quantum Field Theory
 7.Peat,F.David .Superstrings and the Search for the Theory of Everything.
8.Kay, David C. Tensor Calculusp.129 Osculating Plane

9.Kaku, Michio. Strings,Conformal Fields and M theory.

10.Peebles,P.J.E .Principles of Physical Cosmology.p

11Chang,Alan. HAMILTON JACOBI EQUATIONS UNIVERSITY OF CHICAGO 2013.Zeno's paradox: The Math Forum at Drexel University

12:Tipler,Frank j.The Physics of Immortality

13.Godel,Kurt.Godel's Incompleteness Theorems en.m.wikipedia.org

16:ibid item#7 p.237-42

14.Randall,Lisa. Warped Passages
 15:Green,BrianThe Elegant Universe.
and
14.Wick,Mitchell Albert .Megaphysics ,A New Look at the
Universe.

15:Kay,David.CTensor Calculus.
18:Kaku,Michio. Strings, Conformal Fields and M Theory.
19:Wikipedia.Electronen.wikipedia.org/wiki/Electron
20:Spooky Action at a Distance Quantum Entanglement
Wikipedia. Or en.wikipedia.org/wiki/Quantum entanglement
 21:Hau,Len. Harvard Research circa 2003.
BIBLIOGRAPHY

Barrero,John D.The Anthropic Cosmological Principle.Oxford
England.Oxford Press.1986
Brade,Gregg.fractal Time 2009 Library of Congress.
Greene,Brian.The Elegant Universe.NewYork.Vintage Books
editor Random Press.1999
Hawking,Steven and Penrose,Roger.The Nature of Space and
TimePrinceton,N.J;Princeton Science Library 1996
Kaku,Michio.Quantum Field Theory.A Modern Introduction.Oxford
university Press.1993
Kaku,Michio.Strings,Conformal Fiels,and M theory 2nd
edition.Springer Press.2000.
Kay,David C.Tensor Calulus Schaum's Outline Series.
N.Y.McGraw Hill 1998.
Peat,F.David.Superstrings and the Search for the Theory of
Everything.Chicago.Contemporary Books 1998
Peebles,P.J.E.Principles of Physical Cosmology.Princeton Series
in Physics.Princeton University Press 1993
Wald,Robert m.General Relativity.Chicago,Illinose.University of
Chicago Press 1984
Wikipedia: on lin encyclopedia.
Randall, .Lisa .Warped Passages HarperCollins
Publishers.N.Y.2005
Tipler ,Frank J. Physics and Immortality. Anchor Books division
of Random House.1993

16:ibid item#7 p.237-42

16:ibid item#7 p.237-42

Antiparticles, Dark Energy ,and the Big Bang   by Dr. Mitchell Albert Wick

Antimatter was first discovered in 1932 and since then myriad antiparticles have

been discovered including the positron, anti protons, and antiquarks. Antiparticles

are being isolated at Cern, Switzerland in the Hadron Collider but as of this date a

mass of the antiparticle has not been conclusively discovered as positive. It has been

noted that antiparticles have the opposite charge to matter particles but it has not

yet been ascertained that antiparticles have a positive gravity. If antiparticles

mutually repel with antigravity instead of attract with gravity a plausible

mechanism for the 'Big Bang' can be made for the quantum bubble, At Planck Time

10-43 seconds a 50:50 mix of matter and antimatter caused a symmetrical 360

degree(or 2 pi radians)orb blast because of the mutually repulsive force of

antiparticles which forced the explosion converting 99.9999% of the antimatter into

Dark Energy via the formula E=mc2 postulated by Einstein where m=mass of the

antimatter. Whether the mass is positive or negative is up to speculation but since

the Law of Conservation of Energy was purportedly violated by the 'Big Bang' with

antimatter having a positive mass it can be concluded by induction that the mass is negative which is why so little antimatter exists today. As the antiparticles repel

dark energy pushed outward in all directions carrying the balance of matter with it

which it subsequently attracted to other matter by gravity but   (kl)not

to antimatter. The formula (my derivation is included in this work)R jikl- Rji

=-8p[(Geji)=g ji(in four dimensions of space-time reflects the magnitude of dark

energy's repulsive force as curvature of space-time(opposite curvature to

gravity)(gravity was described by Einstein as space-time curved by mass)and the

right side reflects the mutually repulsive force of antiparticles and g ji is the metric

of the antiparticle where i=initial event and j=final event.- 8 pi G where G=the

gravitational constant is from the Cosmologic Constant of Albert Einstein which

reflects the mutually expanding repulsive force pushing galaxies father and farther

apart purportedly fro Dark Energy e=2.71828 and e ji is the vector product of e

from j to i.

This assumes a symmetrical 360 degree Orb Blast in the 'Big Bang' and R is the anti-

gravitational effect on particles and R ji is the antigravity effect on antiparticles. In a symmetric orb blast of 360 degrees or 2 pi radians there is an angle of trajectory of

180 degrees or pi radians. The cosine of pi radians=-1 which explains the -8(pi)G on

the right side of the formula. There are an infinite number of 180 degree slices in a

perfect sphere so the angle of trajectory for an isotropic universe must be 180

degrees or pi radians.

Above formula   Rjikl-Rji=-8(pi)G e ji=g ji or -8(pi)G e(ij,ji)= g ji or R jikl-Rji =-

8(pi)G e(ij,ji)=g ji

1/Rij^2 =space-time of the cosmologic constant or Dark Energy Rij'R ji^2 as the vector

product reveals e ji R2 ij R ji/THE ABSOLUTE VALUE OF R SQUARED ij times the

absolute value of R ji where R 2ij R ji/the absolute value of R SQUARED ij times the

absolute value of R ji =cos theta and theta is pi radians domain 0, or equal to theta ,

or equal to pi range cos theta -1 , or equal to cos theta , or equal to 1 theta is pi

radians. As the trajectory for an isotropic universe is pi radians an infinite number

of 180 degree slices are in a perfect sphere The cosine of pi=-1 g ji=metric for an

antiparticle g ij=metric for the matter particle. R ji=1/R ij 2=1/8(pi)G .g ji where R ji is antigravity for the antiparticle1/ R ij squared is spacetime of the cosmoloigic

constant 1/ 8(pi)G   is the cosmologic constant and g ji is the metric of the anti-

particle.The vector product R ij'R ji=e (ij,ji)=cos (pi)= - e(ij,ji) R jikl-R ji kl=-e(ij,ji)= g

ji/8(pi)G where R jikl= covaruiant t ensor for space-time curvature from

antiparticle of metric g ji on antiparticles R kl=contravariant tensor kl with

covariant tensor ji ji from antiparticle g ji kl is from gravity of contravariant tensor

from metric g ij for matter.   K

                ji

        ijkl
Note R =antigravity affect on particles from antiparticles , The -8(pi)G reflects the

anti-gravitational moment of antiparticle anti-particle interaction

        kl
Rjikl-Rji=-e(ij,ji)=-8(pi)G e(ij,ji)=g ji   Q.E.D.

DARK ENERGY,ENTANGEMENT AND THE COSMOLOGIC CONSTANT

BY Dr . Mitchell Albert Wick

DARK ENERGY , ENTANGLEMENT ,AND THE COSMOLOGIC

CONSTANT

M Theory is the compilation of the five types of string theory. They are based on the

equation

(Mass)^2=2π *Tension where the tension of a string is the radius of the circle.*

The mass is the square root of 2 pi(tension on each string)(total number of strings)

and the mass is the mass of our universe or 10^54kg while the tension on each

string (or potential energy)=10^39 tons. Making the calculation of tons to kilograms

gives the number of strings and the potential energy or V(x) where x =one string can

be put into the Schrodinger Equation to get the wave function. $i\hbar\,\dfrac{\partial\Psi}{\partial time} = \mathcal{H}\,V(x) +$

$V(x)\Psi(x)$ *and* $\mathcal{H} =$

Hamiltonian Operator operating on potential energy or the Tension on each string or $10^{39}tons$

M theory condenses everything into membranes with the D-0 brane as being the

infinite zero dimensional plane 1Brane being the monopole or electron(positron)

2Brane being electromagnetism ,3 Brane being flat matter incorporating strings

acting upon electromagnetism which is in 4 and 5 space as the 4 brane and 5 brane respectively. The number of membranes depend on the number of quantum energy levels and dimensions of space and may in all actuality be over the 26 non-compactified dimensions of string theory. The equation mass of a string)^2=2piTension of a string (or Potential Energy)can be utilized to determine space-time or the circumference of the circle of everything as space-time=space/mass so mass=space/space-time and the square root of 2 pi Tension of 10^39 tons of a string or potential energy =1/time or times arrow is reversed as in the case of the tachyon where the mass squared goes backward in time. The reciprocal of time brings about reciprocal space-time curvature as in the case of anti-gravity as in Dark Energy which is the major component of the mass of our universe pushing everything apart from itself causing a lack of cohesion down to the fermions and even space if Dark Energy approaches infinity. The left side of the Schrodinger Equation reflects Momentum of the wave function and the right side reflects potential energy.

New Evidence that a Multi-verse Exists

The new James Webb Telescope is anticipated to bring conclusive(hard) proof that

Steven Hawking's postulate of a multi-verse in the infinite worlds hypothesis of Lisa

Randall is true.   This is important as it shows that the First Event is the Time

Oscillation Paradox and that "The Big Bang" is only one of a string of Big Bounces

between universe and universe(although the term is a misnomer). There could be

close to a googolplex of universes in the Multiverse as postulated in this author's

previous books" Megaphysics III;Nothing Doesn't Exist" and "The Nth Power "as

well as being mentioned in subsequent publications.Previously only a shifting of

what U.S.C. physicists called shifts in Dark Flow as perturbations were only

suggestive of a multi-verse. Now the relationship of entropy, symmetry and shifts in

charges while universal as far as physical laws are concerned in every system being

observed these shifts may indicate that these laws are shuffled in a different way or

format with different observable reference frames in other universes. The idea of

isotropy(observational symmetry) may be altered as the measuring device of our

universe may be outside of our universe showing the true center of mass or gravity of our universe as being the site of the Big Bang or multiple primordial black holes

which may have contained most or all of the dark matter now being used to "glue

together"ordinary matter into cohesive structures unlike anti-matter which rips or

repels the cohesiveness of structures.

MATHEMATICAL   PROOF   THAT THE HIGGS' FIELD IMPLIES THE EXISTANCE OF
GOD   by Dr. Mitchell Albert Wick

Visualize a vortex with 2 disparate component fields in solenoids. The fields can be
called
ω and ω'. If the field described by ω' has a significant gradient to ω and ω has a chosen

Volume whereby
ω is outside and ω' is inside there is a significant gradient for the harmonic function

Between the inner and outer volume. For all

ω and ω' satisfied if at the boundary of regions (R) the derivative of the function

 The Faber-Jackson Relation =normal projection of the vorticies. Using the Tensor

Virial Theorem and the vortex bulge systems with the Faber-Jackson Relation

Re^$\alpha \propto$

$\sigma(0)^2$ which is the region in topologic space of the exponential function of σ (space)

2 dimensions are zero leaving an infinite number of parallel planes whereby
$MH \rightarrow \sigma$ and $0 \rightarrow M$.

H=Hamiltonian Operator $=\hbar \dfrac{}{2m} \quad \nabla \;^\wedge n \quad n \rightarrow 0 \quad m = mass$
$\nabla = LaPlacean\ Operator$

$MH \propto \sigma\ 0^4$ whereby $4 = \#dimensions$.

The anti-symmetric tensor field of spiral spacetime interacts with the massive Higgs

Field and massless Permanent semi-radius tachyons via the LaGrange Identity and

LeGendre Transformation using the action formula for strings interacting with the

Massive Higgs Field. The Abelian Higgs Model for Abelian Higgs Vorticies is anti-

symmetric equal but opposite cancelling via the Gauss-Bonnett Identities such that

$$\Gamma.\nabla x)\Gamma = \nabla x.\Gamma)\Gamma = -(\Gamma.\Delta x)\Gamma \ where \nabla = harmonic \ function$$

$$\Gamma = spherical \ boundary \ index$$

The boundaries of harmonic tensor fields relate to space-time vorticies and the

Higgs Field. Anti-symmetric or skew symmetric tensors involved in Riemannian

Space-time A^t=-A and a ij=-a ji via Bianchi Identity for space-time from I to j. For

all$\forall$ i to j

i=initial event and j the final event

The le Grange Identity is g det(g I j)> 0.

Spacetime coordinates ds/dt)^2 as in the line element has ds=ds/dt . g I j dx^ i dy
j=IType equation here.

l^ j≤ i

The manifolds or surfaces of spacetime can translate into tensor fields so ds^2= g i j. dx ^ i dy ^ j

And (x+y)y'+ x –y =0

g ij = r i r j

g= l j μ^i μ ^ j

The gradient volume inside the Harmonic Function as nabla approaches 0 or the

Cosmologic Constant $\nabla x \to 0$ so $\Gamma . \nabla x) \Gamma \to 0$ and as antisymmetric tensor fields

$(\Gamma . \nabla x) \Gamma \to 0$ or Λ so $- (\Gamma . \nabla x) \Gamma \to 0 = \Lambda$ .If -
$(\Gamma . \nabla x) \Gamma = -g$ ji and $(\Gamma . \nabla x) = g i j$

In an anti-symmetric abelian tensor filed g ij = g ji

G ij=r1r2 and g=det(g ij) then (g ij)^2= g i^2 j^2- g(i j) where g I j = volume of

the vortex and as (vortex

$\Gamma . \nabla x) \Gamma \to 0 = \Lambda$ and $- (\Gamma \nabla x) \Gamma \to$

0 any vortex including space $-$ time of a harmonic function of ∇ x where $\Gamma =$

index of the spherical boundary or interface as 0 volume is approached equaling the cosmolo

HOW WILL THE SPACETIME CONTINUUM AND OUR UNIVERSE IN THE MULTIVERSE END?

By Dr. Mitchell Albert Wick

As was stated previously in the First Event being a Time Oscillation Paradox and re-schuffling of an infinite number of complete parallel planes or D-0-branes in the spiral space-time continuum the time preceding this event had parallel planes which didn't touch each other due to the repulsive force of the ground state energy limit or Cosmologic Constant. As there were an infinite number of planes each separated by 10^-55 joules in the Cosmologic Constant this totaled as Dark Energy with almost an infinite amount of anti-gravitational or repulsive power which now propels our galaxies away from each other at an exceedingly accelerating rate making the Hubble Expansion Coefficient needing to be changed. Dark Matter acts as Cosmic Glue holding matter together during this expansion but eventually the repulsive force of Dark Energy will break Dark Matter apart and force matter into quarks then gluons and leptons becoming what approaches to an evenly distributed soup of fermions in infinite space-time containing an indescribably huge amount of Dark

Energy.

 As the space-time continuum was formed by the First Event the space-time

continuum will end in a stasis of fermions and tachyons smoothed over Dark Energy

like Jam on bread containing all the matter of the multi-verse and approaching

infinite energy being Dark Energy. Time will still exist but sped up such that a

second of our time will become faster than a nano-second or even a pico-second in

constricted time moving infinitely fast as in a vacuum state and the speed of light

barrier will be broken by the Dark Energy allow the leptons(gluons and quarks will

eventually break)and tachyons to mix forming another Time Oscillation and

forming a new cycle as the contents of the spiral space-time continuum will be

mashed up and broken apart.

Ramajian Effect of Infinite Sums

According to Ramajian the infinite sum from 1 to infinity is $-1/12$. $\sum_1^\infty 1/\sqrt{N}$ is

such that $1/\sqrt{N}$ is $-1/12$ which is a constant. The anti-derivative of a constant is

one as $-1/12 \int_\Lambda^\infty 1/\sqrt{N} = -\frac{1}{12}$ .Based on this in N eigen-states of energy the

factor of entanglement is $-1/12$. The derivative of a constant with regard to a

variable is the constant times 1 which is dx/dx. As a conclusion

$\Gamma\, abcd\ with\ regard\ to\ entanglement's\, effect\ on\ space - time\ is -$

$\frac{1}{12}$ $incorporating\ the\ Boltzmann\ Equation\ for\ N$

This fits in with the mass of dark matter being mass/i   and space-time with regard
to dark matter being space(time/i)   and as mass/i(mass/i)=-1 this fits with
Ramajian's infinite sum equation in space-time which incorporates imaginary
numbers showing the heterotic property of flipping dimensions. Please note that
$\sum_{-\infty}^\infty \sim -$
$\frac{1}{12}$ $as\ it\ brackets\ zero\ from\ both\ opposite\ directions\ \ however 1 \sum_1^\infty Rayo's Number$

Rather than $-1/12$ as Ramajian stated.

WICK'S UNIVERSAL LAW OF MASS AS ILLUSTRATED BY GRAVITY WAVES OR

SPACETIME CURVATURE MEASUREMENTS

When an atomic clock or cesium clock is moved from sealevel to the height of

Mt.Everest time speeds up by approximately two seconds or is constricted slightly.

It speeds up because the distance between the clock and the mass of the earth

increases while the distane of the clock to the mass of the moon decreases so time

speeds up more from being distant to the earth than it slows down or dilates as the

distance to the moon decreases. Also as time speeds up space dilates or space-time

flattens as a vcuum is approached. As we move away from the earth time speeds up

and space dilates or space-time flattens. As we move towards the moon space

curves inward or constricts and time dilates or slows down. In general as a heavy

mass is approached time dilates or slows down and space constricts. As a vacuum is

approaches time speeds up and space flattens or widens. As the observer

approaches the sun time slows down and space constricts heavily and as the void of deep space is approaches time speeds up considerably and space widens or flattens

with minimal curvature

www.ingramcontent.com/pod-product-compliance
Lightning Source LLC
Chambersburg PA
CBHW061502120726
48001CB00004B/1185